Praise for Leander Kahney's books

Inside Steve's Brain
A *USA Today* Best Business Book of 2008

'A rich, essential read for [fans] to get inside Jobs's head and discover what makes Apple insanely great' *USA Today*

'Enjoyable, well-written, very informative, and, most important, up-to-date . . . A unique approach, about [Jobs], from someone as steeped in Apple's culture and history as Kahney' *CNet*

'A detailed, concept-oriented, blow-by-blow look at Apple's CEO and what makes him tick' *Macworld*

Jony Ive
A *New York Times* bestseller

'An adulating biography of Apple's left-brained wunderkind, whose work continues to revolutionize modern technology' *Kirkus Reviews*

'Kahney takes us inside the creation of these memorable objects' *Wall Street Journal*

Leander Kahney, the bestselling author of *Inside Steve's Brain*, *Jony Ive* and *Cult of Mac*, has covered Apple for more than two decades. He is the former news editor of Wired.com, and now the editor and publisher of CultofMac.com.

Tim Cook

The Genius Who Took
Apple to the Next Level

Leander Kahney

BUSINESS

PENGUIN BUSINESS

UK | USA | Canada | Ireland | Australia
India | New Zealand | South Africa

Penguin Business is part of the Penguin Random House group of companies
whose addresses can be found at global.penguinrandomhouse.com.

First published in the United States of America by Porfolio/Penguin,
an imprint of Penguin Random House LLC 2019
First published in Great Britain by Penguin Business 2019
This edition published 2021

001

Printed and bound in Great Britain by Clays Ltd, Elcograf S.p.A.

The authorized representative in the EEA is Penguin Random House Ireland,
Morrison Chambers, 32 Nassau Street, Dublin D02 YH68

A CIP catalogue record for this book is available from the British Library

ISBN: 978–0–241–34821–5

Follow us on LinkedIn: https://www.linkedin.com/company/penguin-connect/

www.greenpenguin.co.uk

Penguin Random House is committed to a
sustainable future for our business, our readers
and our planet. This book is made from Forest
Stewardship Council® certified paper.

For Traci, Nadine, Milo, Olin, and Lyle

Contents

Introduction

Killing It

Each time a man stands up for an ideal, or acts to improve the lot
of others, or strikes out against injustice, he sends forth a tiny ripple of hope, and crossing each other from a million different centers
of energy and daring, those ripples build a current that can sweep
down the mightiest walls of oppression and resistance.

Robert F. Kennedy

When Tim Cook took over as CEO of Apple in 2011, he had big shoes
to fill. One of the largest, most innovative companies in the world
had just lost its visionary founder. Steve Jobs and the company he co-
founded were beyond iconic, and with him gone, pundits predicted disaster.
With rising competition from Android, and uncertainty about future products, Cook had everything to lose by stepping into the driver's seat.

But the critics were wrong. Fast forward eight years, and under Cook's
leadership, Apple has been absolutely killing it. Since Jobs died, Apple

reached the ultimate milestone, becoming the world's first trillion-dollar company, making it the most valuable corporation in the world. Its stock has nearly tripled. Its cash reserves have more than quadrupled since 2010, to a record $267.2 billion—despite its spending nearly $220 billion in stock buybacks and dividends. For perspective, the U.S. government only has $271 billion cash on hand.

To get an idea of just how enormous Apple is during Tim Cook's CEO tenure, consider that the company made $88.3 billion in revenue and $20 billion in profits in Q1 of 2018, as I'm writing this book. By comparison, Facebook, with more than 2.2 billion active users, made just $40.6 billion over all of 2017. Not to mention that in just those three months, Apple made almost as much as its rival Microsoft—once the biggest company in tech—during the entire year of 2017, at $90 billion.

Cook's Apple is crushing the competition in almost every way:

- The iPhone is the single most successful product of all time. It's a *juggernaut.* Apple has sold more than 1.2 billion iPhones in the ten years since it was introduced—four of those years thanks to Jobs's leadership, then the rest under Cook. Cumulative sales are approaching $1 trillion in revenue alone. While Android may ship more handsets, Apple is by far the revenue leader and taking 80 percent of all the profit in the mobile industry. While Apple sells premium handsets with 30–40 percent profit margins, the rest of the mobile industry fights for the low end of the market, where margins are razor thin. And with the iPhone X and its offspring, Apple's market share continues to grow. The rest of the industry is left to fight over smaller and smaller scraps of profit.

- Apple is also succeeding in computers. Although computers play second fiddle to the iPhone, Apple has recently been growing its PC market share for the first time in decades—and is the only company doing so. PC sales are down 26 percent overall from their peak in 2011. Thanks to tablets and smartphones, the PC market seemed unlikely ever to recover. But since Cook took over, Apple has been steadily growing its slice of the market, from 5 percent in 2011 to about 7 percent today. That may seem like modest gains, but like with the iPhone, Apple is only competing at the high end of the market.

- Apple blew open a whole new industry with wearables. Launched in April 2015, the Apple Watch is the first major product of the Tim Cook era that had no input from Steve Jobs. It is a sleeper hit, with more than forty million Apple Watch wearers and sales up 50 percent quarter over quarter. Apple's watch business is already bigger than Rolex. Apple's AirPods are another hit; the company is expected to sell fifty million–plus AirPods and Beats headphones in 2018. With the new HomePod speaker, Apple's smart audio business could exceed $10 billion annually.

- Apple's service business is also growing astronomically. Responsible for $9.1 billion in Q2 2018, it is Apple's second biggest segment by revenue and is almost as big as the satellite TV company Dish Networks. If it were to stand alone, services would be a Fortune 500 company. Some pundits see its services business, built on sales of music, apps, and digital subscriptions, growing to $50 billion by 2020, which could make it bigger than Mac and iPad combined—and bigger than even Disney or Microsoft.

And perhaps the best is yet to come. Behind the scenes, Apple is rumored to be building a robot car, which, if successful, threatens to disrupt the $2 trillion global automotive industry the same way Apple crushed the mobile phone industry. GM and Ford may end up like Nokia and Motorola.

Defying expectations, Apple is enjoying unprecedented levels of success under Tim Cook's leadership, and looks to have a bright future. Despite fears that there would be a mass exodus of talent after Jobs's death and that the company would be gutted as key players left for rivals, Cook has largely held together the management team he inherited from Jobs, supplementing it with clever, high-profile hires of his own. Not only has he steered Apple through a time of uncertainty following Jobs's death, and grown it beyond belief, but he's also led a cultural revolution within the company. Under Cook, Apple isn't as cutthroat and abrasive as it was before—without undoing the company's core products or increasing profits. While Jobs often set teams against each other—and even individual executives—Cook has favored a more harmonious approach, letting go of a few executives who created conflict and drama while increasing cross-collaboration between previously heavily siloed teams.

Cook believes strongly that companies should have a good strategy coupled with good values. In late 2017, his six core values for running Apple were quietly published in an obscure financial statement, and subsequently were given their own subsections on Apple's website. Though they haven't been publicly identified by Cook or the company in any formal capacity, looking at Cook's leadership style over the last eight years, these six values shed light on him as a leader and provide the foundation for everything he has done at Apple:

- Accessibility: Apple believes accessibility is a fundamental human right and technology should be accessible to everyone.

- Education: Apple believes education is a fundamental human right and a quality education should be available to all.

- Environment: Apple drives environmental responsibility in product design and manufacturing.

- Inclusion and diversity: Apple believes diverse teams make innovation possible.

- Privacy and security: Apple believes privacy is a fundamental human right. Every Apple product is designed from the ground up to protect your privacy and security.

- Supplier responsibility: Apple educates and empowers the people in its supply chain, and helps preserve the environment's most precious resources.

As I wrote this book, it became clear how these core values are the bedrock of Cook's leadership at Apple. You'll read how he first unearthed and then embedded them in the company, from the first day he joined Apple to the present day. We'll explore how he developed these values throughout his life and how they came to underpin the heart and soul of the culture at Apple, by investigating the circumstances with which Cook inherited the top job and just how high the stakes were, and then journeying back to his childhood, early career, and his time at Apple.

As the company settles into its recently completed headquarters, a futuristic spaceship that's among the biggest HQs in Silicon Valley, Apple is poised for its third great act, when it brings computing to previously unconquered industries—medicine, health, fitness, automotive, and the smart home, among others. Cook's tenure at Apple is already the stuff of business legend, and it's high time his contributions to Apple, and the world, are aptly celebrated. After all, he has led Apple to become the world's first trillion-dollar company. What follows is the story of Tim Cook, the quiet genius leading Apple to heady heights.

Tim Cook

Chapter 1

The Death of Steve Jobs

On Sunday, August 11, 2011, Tim Cook got a call that would change his life. When he picked up the phone, Steve Jobs was on the other end, asking him to come to his home in Palo Alto. At the time, Jobs was convalescing from treatment for pancreatic cancer and a recent liver transplant. He had been diagnosed with cancer in 2003, and after initially resisting treatment, he had undergone several increasingly invasive procedures to fight the disease ravaging his body. Cook, surprised by the call, asked when he should come over, and when Jobs replied, "Now," Cook knew it was important. He set off immediately to Jobs's home.

When he arrived, Jobs told Cook that he wanted him to take over as CEO of Apple. The plan was for Jobs to step down as CEO, go into semi-retirement, and become the chairman of Apple's board. Even though Jobs was very sick, both men believed—or at least pretended—that he would be around for a while yet. Though he had been diagnosed several years before, he had lived for many years with the disease, refusing to slow down or step back from Apple. In fact, only a few months earlier, in the spring of 2011, he had told his biographer Walter Isaacson, "There'll be more; I'll get to the next lily pad; I'll outrun the cancer." Always determined, Jobs refused to back down or admit that his illness was serious. And at that time, he truly believed he would survive it.

For both men, Jobs's new appointment as chairman wasn't an honorary title or something to keep the shareholders happy; it was a real, honest-to-goodness job that would allow him to oversee and steer Apple's future direction. As David Pogue, technology writer for the *New York Times* and Yahoo, wrote, "You can bet that as chairman, Mr. Jobs will still be the godfather. He'll still be pulling plenty of strings, feeding his vision to his carefully built team, and weighing in on the company's compass headings." Jobs had already left Apple once—and now that he'd made it into one of the most innovative companies in the world, he wasn't about to do so again.

As Jobs and Cook discussed CEO succession on that momentous day in August, Cook brought up Steve's "godfather" role. The pair chatted about how they'd work together in their new positions, not realizing quite how close to death Steve actually was. "I thought . . . he was going to live a lot longer," said Cook, reflecting back on the conversation. "We got into a whole level of discussion about what would it mean for me to be CEO with him as chairman," he recalled. When Jobs said, "You make all the decisions," Cook suspected something was wrong. Jobs would never have handed over the reins willingly. So Cook "tried to pick something that would incite him," asking questions like, "You mean that if I review an ad, and I like it, it should just run without your okay?" Jobs laughed and said, "Well, I hope you'd at least ask me!" Cook "asked him two or three times, 'Are you sure you want me to do this?'" He was prepared for Jobs to step back in if need be, because he "saw him getting better at that point in time."

Jobs's reply to the question about the ad was revealing. He was famously meddlesome in nature, one of the main reasons why Cook assumed he would continue to oversee Apple, even if Cook was now officially in charge of running the day-to-day—though he had largely been doing this for several years already in his role as COO, while Jobs was still CEO. And despite stepping away from all formal responsibility, Jobs did remain very much a part of the company. Cook kept him involved, going "over [to his house] often during the week, and sometimes on the weekends. Every time I saw him he seemed to be getting better. He felt that way as well." Both Jobs and Apple's PR team continued to deny that he was in ill health—no one would admit that he was close to death. But, "unfortunately, it didn't work out that way," Cook said, and Jobs's death stunned the world only a few months later.

Cook the Cipher

When it came to picking a successor to Jobs, there were rumors that the Apple board was likely to choose someone from outside the company, but this was never actually the case. The board was Jobs's board, sometimes controversially so, and they were always going to accept whomever Jobs picked for the role. Jobs wanted an insider who "got" Apple's culture, and he believed there was no one who fit the bill more perfectly than Cook, the man he had trusted to run Apple in his absence on two previous occasions.

Cook, who had been running Apple behind the scenes for so many years, was Jobs's natural successor, but to many onlookers his ascendance to the CEO position was surprising. No one outside Apple or even inside the company would have considered him a visionary, the type of leader whom Jobs had epitomized and everyone assumed Apple needed. It was widely accepted that after Jobs, the next most visionary person at Apple was not Cook but instead chief designer Jony Ive.

3

After all, no one else had Ive's operational power or experience—he had worked hand in glove with Jobs since the days of the first-generation iMac. Together, the pair had spent a decade and more refashioning Apple into a design-led organization. Ive had a cult status of his own, having been the face of many Apple products in promotional videos. For his design on the iMac, iPod, iPhone, and iPad, Ive had won many high-profile awards, and as a consequence he was well known to the public. In contrast, Cook was a much more shadowy figure. He'd never appeared in any product videos and had presented at Apple's product launches on only a few occasions when Jobs was ill. He had given almost no interviews over his career and had been the subject of only a smattering of magazine articles (none of which he participated in). He was largely unknown.

But although some people thought Ive was in a strong position to succeed Jobs, having been so pivotal to Apple's vision and products, he had no interest in running a business. He wanted to continue designing—at Apple he had every designer's dream job: limitless resources and creative freedom. He wasn't going to sacrifice such a rare and liberating position for the management headaches that inevitably come with running an entire company.

Another possible candidate rumored by outside media pundits was Scott Forstall, an ambitious executive who was then in the role of senior vice president of iOS software. Forstall had climbed the leadership ladder at Apple with high-profile projects like Mac OS X, the software that ran the Macintosh. But his star had really risen with the smashing success of the iPhone, since he'd overseen the development of its software. Forstall had a reputation as a hard-charging and demanding executive and styled himself after Jobs, even driving the same silver Mercedes-Benz SL55 AMG. Bloomberg once referred to Forstall as a "mini-Steve," so for some it was a logical assumption that he was a shoo-in for the next CEO. Apple, ever secretive, made no comment on possible successors.

For most, it was baffling that Apple would replace a visionary leader with someone who was so different in character from Jobs, almost his polar opposite. It's easy now to look at Cook's ascent to the head of the world's biggest tech company as the markings of a new era for Apple, but in 2011 it felt more like an ending than a new chapter.

"Nobody would make Tim Cook CEO," a Silicon Valley investor had told *Fortune*'s Adam Lashinsky a few years earlier in 2008. "That's laughable. They don't need a guy who merely [gets stuff done]. They need a brilliant product guy, and Tim is not that guy. He is an ops guy—at a company where ops is outsourced." This was a harsh analysis, but there was a certain truth to it; to most people, Cook was a blank slate, notable more for what he wasn't than what he was.

But ultimately, this unexpected choice was the best for the company. Cook already had the crucial experience of running Apple and had done so effectively. He had stepped in when Jobs took two leaves of absence, in 2009 and 2011, after his initial pancreatic cancer diagnosis in 2003. While Jobs was away, Cook ran Apple as chief executive, overseeing the company's day-to-day operations. He was so unlike Steve Jobs, but he had run the company successfully twice, so the board clearly felt that he would maintain Apple's long-lasting stability.

They had indicated their faith in Cook before. In 2010, as COO, he had received a hefty $58 million in salary, bonus, and other stock awards. Now, as he transitioned into the CEO role, the Apple board voted to award him with one million restricted stock options. To ensure he'd stay on as CEO for a while, half of them were scheduled to vest in August 2016, five years later. The other half were scheduled to vest after ten years, in August

2021. Apple's board was confident that Tim Cook was the CEO Apple needed.

Jobs Resigns; Cook Is CEO

Less than two weeks after he asked Cook to take over as CEO, Jobs resigned and publicly announced him as his successor. A lot of Apple watchers assumed that Jobs wasn't really leaving and that this change wouldn't have a significant impact on Apple, since Jobs would still be a huge part of the company. He had taken leaves of absence before and had always returned. And after stepping down, he was immediately named chairman of the company, which implied his continued oversight of Apple's future.

But Apple's board was concerned about public opinion—they wanted the world to see what they saw in Cook. He might not have been as beloved a figure as Steve Jobs, but it was important that the public learned to love him for his unique strengths—and to have faith that though it would be different, he would run the company just as well as Jobs had done. An Apple press release announced Jobs's resignation and Cook's ascension to CEO. "The Board has complete confidence that Tim is the right person to be our next CEO," said Art Levinson, chairman of Genentech, on behalf of Apple's board. "Tim's 13 years of service to Apple have been marked by outstanding performance, and he has demonstrated remarkable talent and sound judgment in everything he does."

The same day that Jobs's resignation was announced, August 24, 2011, both the *Wall Street Journal* and AllThingsD's Walt Mossberg cited sources "familiar with the situation" as saying that Jobs would continue to be as active as ever in dictating Apple's product strategy. He wasn't going anywhere; Cook would run Apple operationally, but Jobs would be involved in "developing major future products and strategy." People looked for clues wherever they could find them to prove that Jobs was okay; Jobs wasn't quitting the board of directors at Disney or stepping away from Apple completely—most refused to believe that his health had taken a "sudden worsening." Apple's share price only dipped a little—less than 6 percent. Even the market didn't truly believe he was out of the picture.

Cook accepted the role of CEO, acknowledging that he was going to work within the system that Jobs had established. It couldn't have been less like Jobs's return in 1997. Unlike Jobs, Cook wasn't going to tear down what wasn't working and rebuild; he had been a steady captain in his role as COO and planned to keep the ship on its existing trajectory. Unsurprisingly, he did not immediately announce any major changes that would cause investors or fans concern. He wanted to earn their trust first. Plus, according to one widely reported rumor at the time, Jobs had left a detailed plan for a pipeline of products (rumored to be new iPhones, iPads, and Apple TV) that would extend for the next four years at least. Jobs's influence wasn't going away anytime soon. Any changes Cook implemented would be quiet and behind the scenes, just as his previous contributions to Apple had been. Transitioning from COO to CEO, he became more involved in day-to-day administrative matters, something that Jobs rarely had the patience for. He took a more hands-on approach to promotions and corporate reporting structures. He also increased Apple's focus on education and launched a new charitable matching program. (Jobs, by contrast, had canceled many of Apple's charitable initiatives after taking over as CEO.)

Cook wanted to create a sense of company camaraderie, which was lacking when Jobs was at the helm, so he took to sending more company-wide emails, in which he addressed the Apple employees as "Team." One of his earliest such messages as CEO, in August 2011, struck a reassuring tone:

I am looking forward to the amazing opportunity of serving as CEO of the most innovative company in the world. . . . Steve has been an incredible leader and mentor . . . [and] we are really looking forward to Steve's ongoing guidance and inspiration as our Chairman. I want you to be confident that Apple is not going to change. . . . Steve built a company and culture that is unlike any other in the world and we are going to stay true to that. . . . I am confident our best years lie ahead of us and that together we will continue to make Apple the magical place that it is.

Taking a more hands-on approach to interactions with staff was different from Jobs's style. Cook's first email sparked a trend within the company that helped a new culture to develop under his leadership. His emails and other internal communications, such as town hall meetings, helped the new CEO spread his values throughout the company. He also made a conscious effort to adopt some of the things that Jobs had done to establish a sense of continuity between the two leaders. One neat touch Jobs had employed to make himself more approachable was to have a publicly available email address: steve@apple.com or sjobs@apple.com. Cook continued this tradition, responding personally to some of the hundreds of emails that flocked in following his CEO appointment.

One correspondent, a man named Justin R, wrote to Cook, "Tim, just wanted to wish you the best of luck, and to let you know that there are a lot of us that are excited to see where Apple is going. Oh, one more thing—WAR DAMN EAGLE!" (a reference to the "War Eagle" battle cry of Cook's alma mater, Auburn University). And of course, Cook responded: "Thanks Justin. War Eagle forever!" He wasn't just a boring operations guy—these emails gave the public a taste of his personality and showed them that he was a leader dedicated not only to his company but to his customers as well.

Cook was beginning this smooth transition to permanent CEO as the visionary leader who had come to define Apple moved into his new position as chairman. But unfortunately, Jobs would not remain Apple's chairman for long.

The Death of Steve Jobs

Steve Jobs's death on October 5, 2011, shook the world. Just over a month after Cook took over as CEO, Jobs passed away at the age of fifty-six, eight years after his initial diagnosis of pancreatic cancer. He had defied all odds and lived for almost a decade with a disease that has a one-year survival rate of 20 percent and a five-year survival rate of just 7 percent. For so long, people had believed that Jobs and Apple were indestructible. Apple was the company that always performed the impossible, whether that was a dramatic turnaround from near bankruptcy to astonishing corporate success in the late 1990s, unparalleled engineering feats with the iPod and iPhone, or the reinvention of the music industry with iTunes. This was all due to Jobs's influence. Apple was considered untouchable and its leader had become a mythological figure. Few people, it seemed, had entertained the thought that he would actually die.

Jobs passed away the day after Apple unveiled the iPhone 4S at the Yerba Buena Center for the Arts in San Francisco. The 4S's big new feature was the AI voice assistant Siri, one of the final projects Jobs had been actively involved with at Apple. In the crowd at the conference was an empty seat marked "Reserved" for Jobs. He may have been missing in body, but his presence was felt, and the fact that there was a seat set aside for him was even more poignant, foreshadowing his passing away the very next day.

The news of Jobs's death sent ripples of shock and mourning around the world. Never before had the death of a chief executive affected people so intensely. The reaction to his death was unprecedented—despite his often tyrannical leadership of one of the world's most valuable companies, he had maintained a positive public image. He was beloved. He died a few weeks after the start of the Occupy Wall Street movement—a strike back against wealth inequality and "the 1 percent"—and yet he wasn't considered part of that group. People associated him with the beloved iPhones and iPods they carried with them daily, with the MacBooks and iMacs that gave them access to new, potentially world-changing tools. When he died, even Apple's longtime competitor Microsoft flew its flag at half-mast. President Barack Obama hailed Jobs as "among the greatest of American innovators—brave enough to think different, bold enough to believe he could change the world, and talented enough to do it." And the world agreed.

Apple Stores around the world became shrines to Jobs, plastering their windows with fan-made signs and cards commemorating the CEO who, they felt, was one of them. Flowers and candles littered sidewalks outside. Windows were covered in Post-it notes with heartfelt tributes. At the Apple Store in Palo Alto—Steve's hometown—the Post-it tributes completely covered both windows. This type of public mourning was unheard of for a corporate leader.

The months following Jobs's death might have been a tragic time for Cook and those who knew and loved Jobs, but Apple's products were as popular as ever. The iPhone 4S exceeded the preorder and launch numbers of any previous iPhone, with more than four million selling in its first weekend alone. Preorders of Walter Isaacson's authorized biography of

Jobs, a book that would have sold well at any time, also increased by an enormous 42,000 percent on Amazon following Jobs's death.

Running Steve Jobs's Company

As Steve Jobs was being immortalized in every newspaper, magazine, and blog, and on every TV channel and radio station across the globe, the gaze of the world quickly turned to Tim Cook. Doubts about the new CEO persisted while glowing obituaries for Jobs continued to flow. Pundits were skeptical about the sort of company Apple would become without its visionary leader, and fans of Apple were fearful for its future. It was clear from the beginning that Cook's anointment as CEO would be both a blessing and a curse. The role of Apple CEO was a once-in-a-lifetime position that most people wouldn't dare dream of attaining, but it was also one of the riskiest jobs in the world. Jobs's choosing Cook to lead the company was a ringing endorsement of his competence and ability, but to follow in Jobs's footsteps under the pressure and scrutiny of the world was a daunting prospect. Running Apple, Cook would be the most visible CEO in the world: a high-wire act.

It was a scary moment for Cook. While he'd been at Apple for more than a decade, and had risen to become Jobs's most senior lieutenant as chief operating officer, he was now facing a formidable task: taking the reins of an iconic company with millions of rabid fans and a central place in American business and culture. Apple was one of the fastest-growing companies in the world, with a huge, sprawling operation, but it was also facing increasing competition in the burgeoning mobile computing revolution that was sweeping the globe.

The stakes for Tim Cook were higher than ever.

Apple Is Doomed

An intensely private and soft-spoken man, Cook never thought he'd be made CEO. And he certainly never thought he'd replace Jobs. He once famously said, "Come on, replace Steve? No. He's irreplaceable. That's something people have to get over. I see Steve there with gray hair in his 70s, long after I'm retired." Of course, that's not how things worked out.

At the time of his death, Jobs had become modern America's most lionized CEO. Not only had he saved Apple from certain death in the late 1990s, but he'd transformed the company into a massive hit-making machine. The epoch-defining Mac, iPod, iPhone, and iPad had transformed Apple into one of the biggest companies in tech, and certainly the most copied.

Cook had everything to lose. Apple was in danger of losing its leadership position in the marketplace with intense competition from Android, and many thought the company was doomed without its visionary leader. Nobody knew how Cook would act as permanent CEO, since he'd never truly been a public figure.

Cook's reputation initially worked against him—he was certainly a master of operations, but many thought him to be a colorless, unimaginative drone. He had none of the charisma and driving personality of his former boss, which was what people had grown to expect from Apple's CEO. Worse, he didn't have Jobs's imagination. Where would Apple's next generation of insanely great products come from? Jobs had been instrumental in making Apple's products huge successes, and experts in the field were afraid that without him, Apple's run of hits would come to an end.

Even before Jobs officially stepped down, pundits weren't afraid to point out that without Steve at the helm, Apple was doomed. It was no exaggeration: "Why Apple Is Doomed" was the title of a May 2011 editorial in the *Huffington Post*. In it, Ty Fujimura predicted that Apple would never recover from Jobs's death. His "management, even his vision," Fujimura wrote, "is replaceable. But that brilliant sense of taste, to which Apple owes their success, will not be matched by the next regime. His death would leave Apple closer to the pack than ever. . . . Without vastly superior products, their arrogant marketing will fall on deaf ears. Consumers will consider alternatives more readily."

Many others agreed. Jobs was such a singular leader, and Apple's products so closely tied to him, that imagining Apple without him was next to impossible. George F. Colony, CEO of research and advisory firm Forrester, predicted the company would fail without him. "When Steve Jobs departed, he took three things with him: 1) singular charismatic leadership that bound the company together and elicited extraordinary performance from its people; 2) the ability to take big risks; and 3) an unparalleled ability to envision and design products." Apple's momentum, Colony suggested, would only keep it at the top for two to four more years at the most. "Without the arrival of a new charismatic leader it will move from being a great company to being a good company, with a commensurate step down in revenue growth and product innovation."

Cook was not the charismatic leader everyone wanted. He was so unlike Steve Jobs that many analysts, including Colony, drew comparisons with Sony after the departure of its legendary cofounder Akio Morita, Polaroid after Edwin Land, Disney in the twenty years after Walt Disney's death, and even Apple itself after Jobs's first departure in the mid-1980s. The history books are full of companies that stumbled after the death or departure of a crucial founder or leader. Both Ford and Walmart had taken similar dips. Apple's great rival, Microsoft, struggled under the leadership of Steve Ballmer, who took over from the legendary Bill Gates.

Even years later, people continued to doubt that Apple would survive under Cook. "The question of whether Cook can sustain Apple's momentum comes up more often than just about any other question," Michael Useem, Wharton management professor and director of the school's Center for Leadership and Change Management, told *Fortune* magazine in March 2015, three and a half years after Jobs's death. So widespread was the gloom that one of the most hyped books in 2014, three years after Jobs's death, was *Haunted Empire,* by *Wall Street Journal* reporter Yukari Kane, which described Apple as a company anguished by the absence of its former leader. One passage read, "Even as he took control of Apple's sprawling empire, Tim Cook could not escape his boss's shadow. The question was, how would Cook leave that shadow behind? How could anyone compete with a visionary so brilliant and unforgettable that not even death could make him go away?"

Jobs had a vision for Apple that many were afraid would be lost with Cook at the helm. In a 1985 interview with *Playboy* magazine—ironically the same year he was booted out of Apple for a decade—he bemoaned that "companies, as they grow to become multi-billion-dollar entities, somehow lose their vision." At the time of his death, Apple had become a multibillion-dollar company. It was by just about every conceivable metric more successful than it had ever been in its history. But with Jobs as its leader, the vision was still intact. Did Cook have the right insight into and passion for the products, and did he have a vision for the future of Apple?

Those who worked with Cook knew how great a responsibility the former COO was taking on, and some were nervous at first. It was "a daunting challenge," said Greg Joswiak, Apple's vice president of worldwide product marketing, who has worked at Apple for more than thirty years—twenty years as a colleague of Cook's. "It was like, you're riding a bike and it's not just a bike, it's a motorcycle, it's a Harley," he said, in a personal interview at Apple's new spaceship HQ on March 19, 2018. "The challenge was significant."

But if Cook was uneasy about taking on this challenge, it wasn't apparent, even to his closest colleagues like Joswiak. "The world was nervous," but "if [Cook] was [nervous], he certainly didn't show it." If not for his cool demeanor in the face of this significant challenge, Apple would have been a much more difficult place to work after Jobs's death. But Apple employees understood how Cook operated, even if the rest of the world did not. "He took a lot of unfair criticism early on. . . . The outside world wanted to compare him to Steve." But Cook "wasn't going to try to be Steve," Joswiak said. "And what a smart thing because no one could be Steve. . . . Instead Tim was Tim. Tim brought the things that he could to the business."

Like most successful leaders, Cook played to his unique strengths to run the company effectively. In a September 2014 interview with Charlie Rose, he explained that Jobs never expected him to lead Apple in the same way that he had. "He knew, when he chose me, that I wasn't like him, that I'm not a carbon copy of him," Cook told Rose. "And so he obviously thought through that deeply, about who he wanted to lead Apple. I have always felt the responsibility of that." Cook says he desperately wanted to continue Jobs's legacy and "pour every ounce that I had in myself into the company," but he never had the objective of being the same as Jobs. "I knew, the only person I can be is the person I am," he continued. "I've tried to be the best Tim Cook I can be."

And that's exactly what he's done.

Chapter 2

A Worldview Shaped
by the Deep South

Sweet Home Alabama

Timothy Donald Cook was born on November 1, 1960, in Mobile, Alabama, a port city on the Gulf coast and the state's third-biggest city. He was the second of three sons born to Don and Geraldine Cook. Both of his parents were rural Alabama natives. Don worked at the shipyard for Alabama Dry Dock and Shipbuilding, the largest employer in Mobile at the time, building and repairing military ships on Pinto Island. Geraldine worked part-time as a pharmacist and dedicated the rest of her time to the home.

Growing up, Cook had an excellent relationship with both of his parents, and he has remained deeply devoted to them. "He calls every Sunday, no matter what, no matter where he's at," Cook's father told a television interviewer in 2009, two years before Cook assumed the role of CEO at Apple. "Europe, Asia, no matter where he's at, he calls his mother every Sunday. He don't miss a one." Geraldine passed away at the age of seventy-seven in 2015, but Tim still maintains a close relationship with his father to this day.

At some point, the Cook family briefly moved from Mobile to Pensacola, Florida, about an hour's drive away, where Don found work at a huge naval base. But in 1971, when Tim was in middle school, the family returned to Alabama, settling on East Silverhill Avenue in Robertsdale, a small rural town located smack in the middle of Baldwin County, the state's largest county by area. Don and Geraldine chose to settle in Robertsdale so that their three sons could benefit from the top-notch public school system.

School Days

Robertsdale is the picture of small-town southern America. While technically a city, it has an area of only five square miles and a population today of barely more than five thousand, a tenth the size of Cupertino. When Cook was growing up, it had just half that number, about twenty-three hundred residents. Everyone knew everyone in town.

The city had sprung into quiet, laid-back life early in the twentieth century, thanks to its fertile farmland. Agriculture was its main source of income, though later the town would benefit from people passing through en route to the Gulf Shores beaches forty minutes away. When Cook was growing up, Robertsdale was a simple town that had no movie theater or bowling alley, where the most exciting event was the fall Baldwin County Fair. Geraldine described it, not unfondly, as "just a little hole in the ground." The town has had the same mayor for the last thirty years.

The Cook family was religious, leading Tim to become religious himself. He has made references to his Christian belief throughout his career. "As a child, I was baptized in a Baptist church, and faith has always been an important part of my life," he wrote in a 2015 editorial for the *Washington Post*. One can assume that this faith has contributed to his persona as a kind and generous leader. And when he announced he was gay in a *Bloomberg* editorial published in 2014, he referenced God, writing, "I consider being gay among the greatest gifts God has given me." Though Cook is not very outspoken about his religious beliefs these days, it's clear they played a huge part in making him the man and leader he is today.

By all accounts, Cook appears to have fit well into Robertsdale life as a child. Contrary to the teenage tearaway Jobs sometimes portrayed himself as, Cook was a modest, high-achieving, straight-A student. Photos of him from the time depict a slightly gawky but athletic youth with a Donny Osmond haircut and a relaxed, open smile.

Cook excelled at subjects like algebra, geometry, and trigonometry—anything with an analytical edge. In all six years of middle and high school, he was voted the "most studious," and in 1978 he earned the second-highest grades of his year, becoming salutatorian of his graduating class.

His former math teacher Barbara Davis recalled, "He was a reliable kid. He was always meticulous with his work, so I knew it would be done right." Many colleagues and former bosses said the same thing: He could always be trusted to do a job well. And doing things right would become a hallmark of his career.

As well as being studious, Cook was social and well liked among his peers. "You didn't go around calling him a nerd," Davis said. "He was just the kind of person you liked to be around." Many of his peers commented on his intelligence and warm personality, noting he had a playful side, too. Teresa Prochaska Huntsman, the one student who scored higher grades than Cook (and thus the class valedictorian), said of him, "He wasn't one-dimensional. I didn't know anybody who didn't like him. He had a great personality."

Another former classmate and friend, Clarissa Bradstock, said, "He was just really smart, bookish, and had a great sense of humor. We would just hang out. We'd watch *Saturday Night Live* . . . and just talk about school and everything." She added, "It's amazing that somebody from a small town in south Alabama can accomplish what he's done. It's a testament to our country . . . but it's also a testament to him." It's clear that Cook's high school friends are proud of what he's achieved.

Early Business Experience

Cook was a stellar student, but he also excelled in extracurricular activities and showed glimpses of business acumen at an early age. He played the trombone in the school band, and was regularly called on to perform at school dances, football games, parades, and any other local event that required a live orchestra. To earn a bit of cash outside of school, he delivered copies of the Mobile paper, the *Press-Register,* got a job in a restaurant, and worked part-time at the local pharmacy, Lee Drugs, alongside his mother. The drugstore was in the Spaceway Shopping Center, Robertsdale's only strip shopping center, on the main road through town. The drugstore is still there and seems to be thriving, unlike the rest of the strip, which features an assortment of forlorn-looking and shuttered stores. The healthiest businesses are a couple of lunch-counter spots and a tractor rental. Most of Robertsdale's commerce has migrated out to the edge of town, where there's an assortment of fast-food places, a couple of Dollar Generals and Family Dollars and a Walmart Supercenter, where Cook has occasionally been spotted picking up supplies, according to a resident I spoke with when I visited the town in June 2018.

At school, Cook also found time to work on the school yearbook, taking up the very appropriate role of business manager in his senior year. His job was to keep the books and pull in enough advertising to cover the costs for the production of the yearbook. A picture in the yearbook shows the entire production crew in matching sweatshirts (with Cook front and center laughing at something). The shirts are emblazoned with the phrase "Have you got yours?"—presumably a sales tactic encouraging students to buy a copy of the yearbook. That year, thanks to Cook's efforts, the yearbook set new records in the numbers of books sold and ad dollars raised, according to a note in the yearbook itself. Barbara Davis described him as just "the kind of person you need" for a job like that. Working several jobs at once and acting as a business manager on a school project provided Cook with pivotal business experience at a young age, and laid the foundation for the relentless work ethic and sharp business head he was to develop later. Just as he had set new records with the yearbook, he would do the same at Apple many years later.

Other aspects of the yearbook foreshadowed his future at Apple. In one image from his junior yearbook, Cook appears alongside a classmate, showing off a pair of large headphones and an electric typewriter, exciting new technology at the time. The caption reads, "Teresa and Tim are using two of the modern ways to help study." If only he could have known then that one day he'd be leading the biggest technology company ever created.

How Robertsdale Forged Cook's Worldview

Though Robertsdale definitely has the genteel hospitality and old-school charm typical of southern towns, there's also an unpleasant undercurrent of racism running through it. Cook's experience of racism in Robertsdale had a significant impact on him, affecting his view of the world and his future emphasis on equality.

While Cook's parents have claimed they moved from Pensacola to

Robertsdale to get their kids into the best schools possible, the family's relocation coincided with lots of similar moves at the time. Many white families, concerned about the increasing racial tensions in Pensacola as a result of the desegregation of its public schools, moved to nearby Alabama. Although Alabama's public schools had already desegregated in 1963, racial tensions may have been more noticeable in the larger, more racially mixed Pensacola than in the smaller, overwhelmingly white Robertsdale (which is still 85 percent white today, according to the census).

"We had very few African Americans at the school," said Cook's classmate Clarissa Bradstock. "Baldwin County . . . was one of the richest counties at the time, because of oil. But the school we went to . . . was small. I didn't witness any overt racism, but Alabama . . . was still working through a lot of those issues of segregation. I would hear people make racist jokes right in front of people who were African American. That was how it was at the time."

In fact, a few years before the Cooks moved to Robertsdale, the town's central shopping strip had a Piggly Wiggly grocery store that had separate drinking fountains. A Baldwin County resident, who asked to remain anonymous, described the overt racism that he witnessed in the area. "In 1966, my brother dated a young black woman on the sly but was spotted by whites at a café when he stopped to buy the girl a hamburger. She stayed in his car as a precaution. Soon after my brother [who was white] left the café, he was followed and flagged down by these same men in their pickup truck. They hauled him out and beat him badly, leaving him for dead in the nearby woods. . . . He ate soup through a straw for a week." In Alabama at the time, racist acts like this were unfortunately not uncommon.

Cook had his own experiences with racism that would influence him for years to come. In the early 1970s, when he was in junior high, he was riding his new ten-speed bicycle at night along an isolated road outside

Robertsdale when he noticed a fire by the side of the road. He pedaled closer and saw a burning cross, surrounded by Klansmen in white hoods and robes. While Ku Klux Klan membership had diminished from an all-time high of four million in 1925 to just a few thousand in the early 1970s, it was not uncommon to see Klan gatherings in some parts of the South at the time. The Klansmen Cook witnessed had assembled their flaming cross on the property of a local black family. Without thinking, Cook shouted, "Stop!" The Klansmen all looked at him, and one lifted his hood, revealing himself to be a local deacon at one of the churches in Roberts-dale. He quickly warned Cook to keep on moving. It was a shocking experience for the young Cook.

He recalled this experience when receiving an IQLA Lifetime Achievement Award from Auburn University, his alma mater, in 2013. "This image was permanently imprinted in my brain and it would change my life forever," he said. "For me, the cross burning was a symbol of ignorance, of hatred and a fear of anyone different than the majority. I could never understand it." His experience encountering racism would affect the way the young Cook lived his life—and would eventually make it into his business practices.

But despite Cook's insistence on the veracity of his story, there are some Robertsdale residents who take issue with his story about seeing members of the KKK burning crosses in the town. For example, Ted Pratt, a former schoolmate, said, "I have family and friends still in Robertsdale and no one can recall anything like that ever happening. . . . That story really ticked off . . . people who call Robertsdale their hometown." Unsurprisingly, Robertsdale residents past and present do not want to be publicly associated with the Klan, and they are upset that their small town has been painted in such an ugly light by such a high-profile executive.

In a lengthy thread on a Facebook page called "Robertsdale Past and

Present," dozens of Robertsdale residents, current and former, questioned Cook's memories. "Tim Cook is flat out lying about this incident," wrote one commenter named Rod Jerkins. "Never happened." (His comment got half a dozen "likes," and was one of the most liked comments in the thread, showing general agreement with the sentiment.) In fact, there were no comments that I could find in defense of Cook. Among the 143 comments on the post, almost all of them questioned Cook's memory. Another commenter, Marvin Johnson, added, "I asked relatives who have lived here much longer than I have[,] . . . more than half [a] century, and was told this incident never happened." A third commenter wrote, "He flat out lied. Period."

But it's clear from the Facebook thread that many Robertsdale residents are in denial. Public posts on Facebook aren't a good forum for admitting Klan activity, even if it took place decades ago. A couple of people who posted in the group said that while they didn't see burning crosses, they did see *burned-out* crosses, and one recalled a burning cross displayed during a Christmas parade in a nearby town. The bitter truth is that Robertsdale likely had its fair share of Klan activity.

Representative Patricia Todd said that not only was the Klan active during Cook's childhood, but it's also still very much around today. "They've been distributing flyers in the last two years in various communities in Birmingham," she said. "People can't deny Alabama's history around civil rights, which is not good . . . but racism is still alive and well in Alabama. . . . People don't say it out loud, but there's still a lot of people who really hate people who are different than them."

Years after his experience with the Klan, Cook had another meaningful interaction with racism. When he was sixteen, he won an essay contest organized by a local utility, the Alabama Rural Electric Association. The topic was "Rural Electric Cooperatives—Challengers of Yesterday, Today, Tomorrow," and Cook wrote his essay by hand; his family couldn't afford a typewriter.

His prize was an all-expenses-paid trip to the nation's capital, where he attended impressive banquets and got to hear President Jimmy Carter speak at the White House. But the trip was tainted by the fact that Cook also met Alabama governor George Wallace, a staunch segregationist who had, in vain, resisted the federal government's attempts to integrate the state's public schools during the 1960s. Cook shook Wallace's hand, but later regretted it. "Meeting my governor was not an honor for me," he said. "Shaking his hand felt like a betrayal of my own beliefs. It felt wrong, like I was selling a piece of my soul." Luckily, he was able to learn from this experience—today, he has absolutely no tolerance for racism in any form, and though it's still a work in progress, he has worked to make Apple a more inclusive place. Under his guidance, Apple has hired a greater proportion of minority workers than many of its Silicon Valley peers and given generous grants to historically black colleges and charities and foundations to encourage minority students to study STEM subjects.

Many of the values he has implemented at Apple seem to be a direct result of his childhood experiences with discrimination. In a 2013 talk to students at Duke's Fuqua School of Business (where he earned his MBA), Cook talked about following the example of two of his heroes growing up: Martin Luther King Jr. and Robert F. Kennedy. "I was born and raised in the South and I saw, over the course of growing up, some of the worst behavior in terms of discrimination that literally would make me sick," he told the students. He admires King and Kennedy because they risked their lives to fight discrimination. "That's why I have three photos in my offices, two are of Kennedy, one is King," he said. "That's the only photos I have in my office. I look at them every day and . . . I think they're incredible role models for all of us. . . . And that's not a political statement, that's just a statement about treating people fairly."

The hatred and discrimination Cook witnessed during his childhood would stick with him throughout his life, influencing the way he acts in life and in business. Lisa Jackson, the first African American woman to head the Environmental Protection Agency, whom Cook hired to steer Apple's environmental efforts in 2013, said Cook's outlook on life was heavily influenced by his childhood experiences in the South. "It's just part of who he is," she said. "Being from the South, you've seen the ugliness but you've also seen the promise and the possibility," she added. "At least I can't divorce it from who I am, and Tim's spoken about it."

Roots of Alabama Activism

Cook has maintained his principles throughout his career, at Apple and elsewhere. In a 2015 commencement address at George Washington University, he expressed his belief that a person should not have to choose between "doing good and doing well." His refusal to compromise his values has contributed directly to Apple's success, though he's been tested many times. In May 2014, a member of a conservative think tank, the National Center for Public Policy Research, pushed him to consider the impact sustainability programs were having on Apple's bottom line. But he refused. "When we work on making our devices accessible by the blind, I don't consider the bloody ROI [return on investment]," he said. And the same thinking applies to Apple's environmental initiatives, worker safety, and other policies. "If you want me to do things only for ROI reasons, you should get out of this stock," he snarled at the conservative investor. Afterward, the NCPPR issued a statement decrying Cook's stance: "After today's meeting, investors can be certain that Apple is wasting untold amounts of shareholder money to combat so-called climate change." But Cook, as always, stayed true to his principles.

This moral compass, developed in his childhood, is one of the biggest differences between Jobs's and Cook's public personas. Jobs eschewed charitable donations, appeared to care very little about sustainability, and rarely spoke out about social issues. For him, the products he was willing into the world were contribution enough. A Macintosh with a gorgeous graphical user interface was more than enough to put a dent in the universe. Jobs believed that his work at Apple was more important to the world than any charitable contribution could be. For Cook, on the other hand, his contribution to society is more nuanced and complex. While he has always spoken with immense pride about the quality of Apple's products, he has also been outspoken about using his position as CEO of the world's most valuable company to make Apple a "force for good." As we'll see in later chapters, Cook has greatly increased Apple's charitable giving, taken significant strides in making the company a major force in renewable energy and ensuring that its products are less toxic and more recyclable, attempted to render Apple's supply chain safer and less exploitative, and made significant efforts to make Apple a more inclusive and diverse workplace.

His morality is rooted in a Christian upbringing, southern manners, and the teachings of his heroes Martin Luther King Jr. and Robert F. Kennedy. "I drew on the moral sense that I'd learned from my parents, and in church, and in my own heart, and that led me on my own journey of discovery," he recalled in one speech. He drew on experiences from books as well. As a kid, he reportedly borrowed a copy of Harper Lee's *To Kill a Mockingbird* from the library in Robertsdale. Its depiction of good-hearted lawyer Atticus Finch's righteous battle against racism in a fictitious Alabama town clearly resonated with him.

Cook's support of marginalized minority groups was influenced too by his experience growing up gay in the South. He had never discussed this aspect of his life publicly, until talk show host Stephen Colbert asked him directly about it in a 2015 television interview: "That experience of growing up in Alabama as sort of a resident outsider because of your sexuality, did that inform in any way your trying to help people who are in hardship around the world?" Cook responded in the affirmative, explaining that he felt he needed to do something to counter widespread homophobia. "Kids were getting bullied in school, kids were getting basically discriminated against, and kids were getting disclaimed by their parents, and I needed to do something," he said. "I felt a tremendous responsibility to do it." This response offered a small, rare glimpse into the personal life of a very private man.

Others have echoed Colbert's line of thinking when it comes to joining the dots on Cook's own homosexuality and his willingness to speak out about progressive human rights issues. "I have to believe that growing up in Alabama during the 1960s and witnessing what he did, especially as someone who is gay, he understood the dangers of remaining silent," said Kerry Kennedy, daughter of Robert F. Kennedy and a human rights activist herself. Now "he's not afraid to stand up when he sees something wrong."

Cook has not shared much about his experience as a young gay man, and by all accounts, it's unlikely he came out in high school. His friend Clarissa Bradstock said she did not know he was gay at the time—and she even had a crush on him. She explains that their high school may not have been the most accepting environment. "There were some other people in the school that I thought at the time probably had a different orientation," she said, "but you never said anything. They didn't get harassed, but of course, they weren't allowed to come out."

"Robertsdale is not exactly a liberal bastion of the state," said Representative Patricia Todd, the state's only openly gay legislator. "If he did come out at a younger age, I would think it would [have been] very difficult." This sentiment was echoed by a Baldwin County resident, who asked to remain anonymous: "Chances are that the locals would have shunned Tim if they had known he was gay. Stories abound of gays being ambushed and beaten, even by cops, who would tell a judge the victim 'fell down the stairs.'" It's no wonder that Cook chose not to make his sexuality public while he was living in Robertsdale.

But now many people are praising Cook for coming out, including Todd. She says it has empowered others to follow his lead and has normalized being gay. "I do think that makes a big difference, people like Tim and other openly gay CEOs," she said. "I think it makes people realize how diverse our community is and that we truly are everywhere. Unlike other social movements on gender or race or ethnicity, we can hide. But now people are feeling empowered to come out. Tim's been a part of that. . . . When the CEO of the most successful company comes out, people take notice." This was exactly Cook's intention. As he wrote in the *Bloomberg* "coming out" editorial, "If hearing that the CEO of Apple is gay can help someone struggling to come to terms with who he or she is, or bring comfort to anyone who feels alone, or inspire people to insist on their equality, then it's worth the trade-off with my own privacy."

Hometown Nonhero

Though many Robertsdale residents know of Cook's achievements, these days he isn't overtly celebrated in his hometown. Some of his old friends and classmates are openly proud of what he's achieved, but there aren't any plaques or trophies celebrating his accomplishments, old or new. A glass case in his old high school commemorates football star Joe Childress, an

NFL running back who graduated in the 1950s, but there's still nothing to be found about Cook. Perhaps it's because he's a businessman rather than an athlete. One current resident said most townsfolk wouldn't know who Cook is, since they're not concerned with global CEOs in their small Alabama town.

Among some who do know who Cook is, there's a feeling that he has not done as much for the local economy as they believe he should have. On the "Robertsdale, Past and Present" Facebook page, there is a lengthy discussion of why he hasn't used Apple to give back to the area. One poster asked why Robertsdale isn't a tech center for Alabama, since it's Cook's hometown. This seems to be a commonly asked question.

Current resident Dillan Gosnay, twenty-one, said the town is struggling for jobs. He works the odd welding job when he can, but he's essentially unemployed. And he's not the only one. Like a lot of rural areas in America, Robertsdale has been struggling with unemployment for a long time. "The jobs, that is the worst part about living here. It's hard, everybody has to work three jobs. . . . There's nothing, there's no big-name companies that make their headquarters here," he said. "I think a lot of people wish that he would bring something and put it here for more jobs."

But it will be awhile before Apple comes to town. Currently, Alabama doesn't have explicit laws barring discrimination on the grounds of race, age, or sexuality. Shortly after Cook was inducted into the Alabama Academy of Honor in 2014, he personally told Representative Todd that Apple had no intention of investing in Alabama until the state passes antidiscrimination laws. "Citizens of Alabama can still be fired based on their sexual orientation," Cook said. "We can't change the past, but we can learn from it, and we can create a different future."

Shortly after and inspired by Cook's visit, Representative Todd introduced a nondiscrimination bill named after him in the Alabama state legislature. "Tim was honored to have it named after him," she said. But unfortunately, it never became law. "Of course, it didn't go anywhere," she said. "I'm a Democrat and the legislature is controlled by Republicans. They have a supermajority. . . . They weren't going to pass a nondiscrimination bill. It was more symbolic than anything. But . . . at least we got the conversation going." It failed, but it was a step in the right direction.

And it's clear that Cook isn't indifferent to his hometown or home state. "He's very interested in what goes on in Alabama," Representative Todd said. "He keeps up with it. . . . He's trying to help the state move forward. Which . . . considering it's Alabama, is going to take a long time." In December 2014, he donated an undisclosed but "sizeable" sum to the Human Rights Campaign, a Washington-based advocacy organization, which launched a three-year, $8.5 million campaign for gay rights in Alabama, Arkansas, and Mississippi. HRC's "Project One America" is ongoing, with field offices and full-time staff members in all three states. Since Cook's donation, HRC has grown to be the largest LGBTQ civil rights organization in the country, the group claims, with more than three million members and supporters. Cook also donated iPads to a very poor school system in Alabama's "Black Belt," an impoverished region named after its distinctive black dirt. "He has contributed, but he's made it very clear to elected officials here, especially the legislature, 'I'm not going to expand my operations in Alabama until you all pass nondiscrimination,'" Representative Todd said. "Of course, we're not going to do that 'cause we'd rather, you know, thump our Bible than bring good jobs to Alabama."

Cook has also inspired the local business community to speak up about nondiscrimination—with the current laws, and divisive efforts like bathroom bills, it's difficult for them to recruit businesses to Alabama. "If you'd told me ten years ago that our biggest advocates for nondiscrimination were going to be the military and corporations, I would have laughed at you, but that's what's happening," Representative Todd said. "It's hard to recruit good employees and economic development when you're seen as a state that's backwards and allows discrimination." Even though progress has been slow, Cook is hopeful that Alabama will change its laws—after all, the place is special to him. He returns regularly to watch Auburn football games and visit his family. He spent the first twenty-one years of his life there, and it clearly had an impact on him. As he once told a group of young people while visiting Birmingham, Alabama, "Most of my formative years were spent in Alabama." Hopefully some of them will follow his lead and change Alabama for the better.

Engineering at Auburn

After graduating from high school in 1978, Cook left Robertsdale to attend Auburn University, where he pursued a bachelor of science degree in industrial engineering, one of his long-term goals. "Ever since he was in the seventh grade, he said, 'I want to go to Auburn,'" his mother recalled. Auburn University was fairly close to Cook's hometown—a mere three hours by car. Staying in Alabama was important to him. The other choice of university was the University of Alabama, in Tuscaloosa, but he felt it was too posh. "The well-to-do people went to Alabama," he explained. "And it was sort of the place for the doctors and the lawyers, and I always associated myself with the working people. And the working people went to Auburn."

His choice of industrial engineering at Auburn was astute. Former Chrysler CEO Lee Iacocca, former Walmart CEO Mike Duke, and former United Parcel Service CEO Michael Eskew all had industrial engineering backgrounds, and earning this degree would put the young Cook on the same path. It suited his sensibilities and his skills: Industrial engineering focuses on ways to optimize complex systems, how best to eliminate wasted expenditure and make the best possible use of resources. This was a skill Cook developed early on. "He could cut through all the junk and get down to the gist of [a] problem very quickly," said one of his professors, Robert Bulfin.

Cook's career at Auburn was solid, if not spectacular. He was nominated an outstanding engineering graduate in his senior year and was incredibly humbled by the accolade. "I don't deserve this," he insisted at the time. "There are any number of people who deserve this more than me."

Saeed Maghsoodloo, one of Cook's college professors, remembered him as a "solid B-plus or A-minus student." In an interview for the *New York Times*, published the year that Cook became Apple CEO, Maghsoodloo painted a picture of Cook as "a very quiet, unassuming individual [who was] very, very intense" and "sat quietly and studied." But he was also personable and a good friend, as popular at university as he was in high school. Photos from the time show him laughing and joking with groups of his friends.

At Auburn, he would learn many of the skills that would help him throughout his career. He learned how to program. For one class, he created a system to improve the timing of traffic lights near the university. "I tried to optimize traffic because at that point in time stop lights were set on timers," he explained. "I wanted to come up with a way to reduce the queues so people didn't have to wait as long, whilst keeping the environment safe." His system apparently worked so well that the local police adopted it. "That was pretty cool at the time—and it worked. Law enforcement implemented it," he said. These days, though, he's lost some of his coding ability. Now he jokes that his coding skills are "not bad," but "there are many, many people in Apple who are better than I am."

Auburn seems to have had a significant impact on Cook's approach to work and his world outlook. The Auburn Creed, written in 1943 by the university's first football coach, George Petrie, states, "I believe this is a practical world and that I can count only on what I earn. Therefore, I believe in work, hard work. I believe in education, which gives me the knowledge to work wisely and trains my mind and my hands to work skillfully. I believe in honesty and truthfulness, without which I cannot win the respect and confidence of my fellow men."

In a commencement speech at Auburn in 2010, Cook echoed these words as a personal mantra. "Though the sentiment is a simple one, there's tremendous dignity and wisdom in these words and they have stood the test of time," he told the audience. "Those who try to achieve success without hard work ultimately deceive themselves, or worse, deceive others." Cook certainly believed in hard work from a young age, and it is clear from the way he runs Apple that he values it in his employees as well.

At Auburn, Cook got his first true experience in corporate management. He enrolled in a cooperative education program, which meant he spent part of his time in college at Reynolds Aluminum in Richmond, Virginia. It turned out to be a crash course in the realities of the working world. Almost immediately, the firm was forced to make a large number of its employees redundant. Their loss, however, was Cook's gain, as he was given the opportunity to fill in and help run the company alongside its president. The role of second in command was one that he would perfect over the following decades before rising to the top at Apple.

Cook graduated from Auburn in 1982, roughly eighteen months after Apple's IPO and eighteen months before it released its breakthrough Macintosh. But Apple wasn't on the twenty-one-year-old's radar yet. When he graduated from college, a recruiter for IBM, the computing powerhouse that had just released its first IBM PC, approached him about a job. He also received offers from Andersen Consulting and General Electric, both attractive companies to work for. On making his decision to join IBM, Cook said, "The truth is, I'd never thought much about computers. Would things have turned out different if that hadn't happened? I don't know. But I do know that there are only a very few things in life that define you and that was one of them for me." Despite his never having considered a job in the tech industry before—the industry was in its infancy and wasn't likely high on many graduates' lists—the IBM job was a good one, and Cook accepted it.

Immediately upon receiving his degree in electrical engineering, Cook joined IBM's fast-growing PC operation, a relatively new division of the computing giant that operated out of a big assembly plant in Research Triangle Park in North Carolina. With the exception of the occasional trip to Auburn for football games, he never looked back.

Chapter 3

Learning the Trade at Big Blue

The IBM PC

Securing a job at IBM was a lucky break for the young Cook. The computer industry was an exciting and booming industry in the early 1980s, and for a person with his talents and drive, the potential rewards were astronomical. It's little surprise that he flourished. The home computer industry was still new, and despite the success of machines from Apple, Atari, and Commodore, and growing interest in IBM's Personal Computer, less than 10 percent of U.S. households owned a PC of their own. The market was at the beginning of a giant explosion, and manufacturers were fighting to attract customers who were shopping for their very first computer.

At this time, IBM was the white-hot center of the industry. Before launching its first PC, IBM had been known for selling massive, room-filling mainframes to corporations and governments. The company already employed more than 350,000 people worldwide, and in 1981 the computing giant had decided to get into the fledgling PC business, largely based on the success of Steve Jobs and Steve Wozniak's Apple II, the little computer that could.

IBM's first Personal Computer was a breakthrough machine that is now considered a landmark: It was a huge commercial hit, and because it was built from standardized parts, it was widely copied. Its name, "Personal Computer"—shortened to PC—became the shorthand for an entire class of small, fast, inexpensive machines. Within a decade, IBM clone machines, as they were called, would dominate the multibillion-dollar PC industry.

The $1,565 Personal Computer ran BASIC, a popular programming language at the time (Microsoft Windows wouldn't launch until a couple of years later, in 1985), and promised flexibility, performance, and ease of use. It packed a 16-bit microprocessor, 16 kilobytes of RAM, and 40 kilobytes of storage, which was huge at the time. But compared to today's computers, which have magnitudes more processing power and storage space, it was slow and primitive. Take, for example, a Series 3 Apple Watch, which has four hundred thousand times more storage space (16 gigabytes versus 40 kilobytes) in a tiny device that fits on your wrist. The PC's "high-resolution" display had room for twenty-five lines, each of which could hold up to eighty characters. One of the first brochures for the Personal Computer, which was the most affordable IBM computer in 1982, boasted about its "advanced design features and a variety of productive program packages which add to the rewards and enjoyment of having a computer of your very own." The machine was especially popular with the business community, which wanted a relatively cheap, flexible computer to run routine office tasks like accounting, communications, and billing. It quickly became the standard in offices around the world. What was once a niche industry for nerdy hobbyists became mainstream in the business world.

By the end of that year, IBM was selling a Personal Computer every minute. The company had originally estimated that it would shift about 250,000 units over five years, but there were times when it built and sold almost as many computers per month, which made the Personal Computer the biggest competitor to the Apple II. In 1982, IBM and its competitors, which had "cloned," or basically copied, the machine, sold 2.8 million units of the home computer in the United States, double the number of units sold in 1981.

Its rapid takeoff earned the computer the honor of *Time* magazine's "Man of the Year" for 1982. "Machine of the Year: The Computer Moves In," read the front cover of the January 3, 1983, issue, which featured a papier-mâché man seated in front of a generic desktop machine. "There are some occasions . . . when the most significant force in a year's news is not a single individual but a process, and a widespread recognition by a whole society that this process is changing the course of all other processes," read the accompanying article. "That is why . . . *TIME* has decided that 1982 is the year of the computer." This was huge not only for IBM but also for the entire technology industry.

It affected Apple, and Steve Jobs specifically. Jobs was upset that he wasn't awarded Man of the Year himself. He had assumed incorrectly that he had secured the title because Apple had become the first computer company to reach $1 billion in annual sales that same year. "They FedExed me the magazine," Jobs revealed to Walter Isaacson, his biographer, in the years prior to his death, "and I remember opening the package, thoroughly expecting to see my mug on the cover, and it was this computer sculpture thing. I thought 'Huh?' And then I read the article." He added, "And it was so awful that I actually cried." It was a major blow to Apple, but this was long before Cook ever thought about working there. He was happily beginning his career working on the IBM PC, which would give him the skills he needed to make a name for himself at Apple.

The Plant at Research Triangle Park

IBM's PC division was based in a big plant at Research Triangle Park. It was growing fast, and it needed lots of fresh recruits. IBM's strategy at the time was to hire lots of college graduates and train and promote from within its own ranks, which is how Cook first ended up there. When he joined IBM, the facility at RTP was a big, sprawling plant occupying six hundred thousand square feet. It wasn't so much a factory as a big assemble-and-test operation. Aside from a few components built in-house (such as keyboards), IBM bought most of the computer's parts from other companies, such as Intel.

The plant operated six production lines that ran twenty-four hours a day, Monday to Friday, in three shifts. It would shut down on weekends unless it was running behind or there were big orders to fulfill. Every single day, as many as a hundred eighteen-wheeler semi-trucks would deliver parts. But there was no warehouse. Parts would come in at one end, and leave a few hours later as assembled computers (at a rate of about one computer per line every minute).

About half the plant's twelve thousand employees were assemblers, and almost all the assembly was done by hand. Gene Addesso, a thirty-six-year IBM veteran and the plant manager at the time, estimated that they assembled six thousand to eight thousand computers a day, rising to ten thousand a day at peak volume. As they were assembled, the machines would proceed down the conveyor belts to a testing area, where they were put through several tests. If they passed, they continued on to packaging, the only highly automated part of the plant. When the process was complete, the packages were loaded on trucks, which would carry them to distributors and customers.

Just-in-Time Manufacturing

The plant was lean and efficient, using a system of just-in-time (JIT) manufacturing, or "continuous-flow manufacturing" (CFM), as it was known at IBM. "Just-in-time manufacturing meant that you didn't have to maintain inventory in a warehouse," Addesso explained. "Make it. Ship it. . . . It saves a lot of time and a lot of money." There was no storing of parts or finished goods, and therefore no need for warehouse space.

The JIT philosophy, commonly referred to as "lean manufacturing" in the United States, was devised to effectively meet customer demand while avoiding a surplus of goods. It was popularized in Japan throughout the 1960s and 1970s—largely by Toyota, which made automation and JIT the two pillars of its entire production system—to make the manufacturing process more efficient and improve the return on investment.

In JIT manufacturing, "a flow process, the right parts needed in assembly reach the assembly line at the time they are needed and only in the amount needed," explained Taiichi Ohno, the former Toyota engineer who was considered the father of the Toyota Production System, in his book *Toyota Production System: Beyond Large-Scale Production*. "A company establishing this flow throughout can approach zero inventory." This was IBM's goal: to minimize investment in parts and computers waiting to be sold, which in the fast-moving PC industry could be outdated in six months.

A common misconception is that Ohno and Toyota gleaned ideas for JIT from the American automotive industry (which practiced a form of

JIT in its own right). The process was actually inspired by the American self-service supermarket phenomenon. Ohno recognized that such mass-merchandising stores restock their inventory to match customer buying habits and frequencies. They stock enough to serve customers with as much as they need, when they need it, then fill their shelves with just enough inventory to continue the cycle. "Sometimes, of course, customers may buy more than they need. In principle, however, the supermarket is a place where we buy according to need," Ohno observed. "Supermarket operators must make certain that customers can buy what they need at any time."

Toyota was a pioneer of JIT, and others quickly followed suit, but references to similar production processes go back a lot further. In his 1923 book *My Life and Work,* Henry Ford wrote, "We have found in buying materials that it is not worthwhile to buy for other than immediate needs. We buy only enough to fit into the plan of production, taking into consideration the state of transportation at the time." He continued, "If transportation were perfect and an even flow of materials could be assured, it would not be necessary to carry any stock whatsoever. The carloads of raw materials would arrive on schedule and in the planned order and amounts, and go from the railway cars into production. That would save a great deal of money, for it would give a very rapid turnover and thus decrease the amount of money tied up in materials."

As developer of one of the first PCs in the world, IBM was a pioneer of JIT for PC manufacturing. IBM first began implementing its CFM program in January 1985 at its Tucson, Arizona, facility, which manufactured optical storage units, laser printers, and printed circuit boards. The cycle time for a printed circuit board operation at this time was 17.5 days, but by December 1987, when CFM was in full swing, that had been almost halved. Quality had also improved significantly as a result, and there was a 100 percent improvement in space utilization. Employee morale had increased 21 percent, too.

Cook's First Job

It was in Cook's first role at IBM that he learned the intricacies of just-in-time, which he would later use to overhaul Apple's entire production process. His first job was on the line at the factory, doing pipeline management—ensuring that the factory had enough parts to make its PCs. And it was trickier than it sounds.

Cook had to ensure that the plant had exactly the right number of parts at any given time for all products that were being made. It was a precise, detail-oriented, and potentially stressful task. Managing the pipeline for this operation was a big, complex job. "It was very difficult because you had a large number of vendors and you always had to have the right sets of parts at the right place at the right time," said Richard Daugherty, vice president and general manager of the RTP facility. "It was a big, big job. If you did it incorrectly, you either couldn't ship product or you got stuck with excessive inventory. Either one could be a killer. And that's the area that I remember Tim working the most in and really had his expertise in."

Everything was computerized and the plant managers knew exactly what was coming when, where, and at what time. Everything was very carefully traced, so that there was just the right amount of inventory in the factory at all times. IBM announced a new PC every six months, putting serious pressure on the inventory teams. When they fell behind, Daugherty said, "I can recall [Cook] and some other folks were responsible for getting it fixed and did." It wasn't long before Cook began to move away from the pack, excelling in pipeline management and operations.

Cook's High Potential

A couple of years after joining IBM, Cook was identified as having high potential, or "HiPo" in IBM jargon. The HiPo program was a big deal at IBM; it was the management track for the company's future leaders. Every year, senior managers at the plant would draw up a list of twenty-five of the most promising young hires. The list ranked things like performance, responsibility, and emerging leadership potential. Cook was ranked number one.

"I had him ranked as number one on my HiPo list," said Daugherty. "So he came to our attention very early as an all-star. And thank God I recognized it and put him up at number one. They just had absolute confidence in him. They trusted him." It was Cook's work ethic, which he had developed through high school and college, that earned him a spot on the HiPo list and ultimately saw him rise through the ranks at IBM. Ray Mays, the former manager for personal computer production at the plant, who was also Cook's boss for a time, agreed that Cook stood out from the pack. "The thing that is impressive about him in my mind is . . . his work ethic," Mays said. "I'm not sure he ever sleeps. In one instance, he was in China and it must have been two or three o'clock in the morning. He responded [to my email] within five minutes. He must have been one of the smartest people that I've ever worked with." Mays added, "Everything that he was and everything he did was thoroughly thought out and executed."

Part of IBM's philosophy was of promoting from within, rather than hiring for leadership positions from outside. Those who made the high-performance list were first in line to be groomed internally for a leadership role. High performers would be moved around the plant, assigned to different departments or tasks to get a more comprehensive idea of IBM's operations. "One of the things we do with folks like that is move them from assignment to assignment to broaden their knowledge of what we do within

IBM," said Mays. As a HiPo, Cook worked temporarily as the manufacturing department manager; at another time he was the administrative assistant to the plant manager.

In addition to managing the plant, Cook and his colleagues would often go to suppliers and check on their operations, making sure they were maintaining the right quality and were able to deliver on time. They would often help suppliers if they had any logistical difficulties. When they came back from these trips, "guess who was chosen to give the presentation to IBM plant management?" said Daugherty. "It was always Cook." Gene Addesso agreed: "You could look at this guy and say this guy's going to be a leader. He can manage people. He just stood out in the group. And the group looked up to him." Though Cook had always been a strong but quiet force in school, at IBM he really started to excel as a leader, and people noticed. He had a natural talent for leadership, and he started to grow into himself at IBM, honing his skills with additional schooling at Duke.

Cook's MBA

All the employees on the HiPo list were sent to colleges to take supplementary management leadership courses. Mays, who had also been a HiPo employee, remembered getting sent to charm school in New York. "It was a public relations operation in New York City that had me there for a week critiquing every move I made [and] how I spoke," he said.

But Cook wanted more and began attending evening classes at Duke University's Fuqua School of Business, which earned him a master of business administration in 1988. Getting an MBA was Cook's idea, but IBM paid for it. "You were kind of foolish not to take advantage of it if you had a desire to get an MBA," said Mays, whose wife, Jenny, also took the MBA program at the same time as Cook. Fuqua's "Evening Executive" program lasted eighteen months and took place at night. It was a tough schedule for Cook and the other MBA candidates. Mays elaborated: "You work all day, and then . . . you spend three to four hours a night [at school], and you have assignments on top of that." But for Cook, it was worth it.

The MBA helped to advance his career at IBM, which needed engineers with management and leadership skills. During a 2016 appearance at the Utah Tech Tour, Cook said he knew that his engineering degree alone wouldn't have been enough—engineers needed to supplement their technical knowledge with a global view of problems worth tackling, which he gained while studying for his MBA. One of the things that impressed colleagues at Apple was his business savvy. "He's a good business mind," said Cook's colleague Greg Joswiak. "Steve kind of demanded that of the leaders. You have that business sense, and Tim had that."

Early Ethics

At Duke, Cook also took an ethics class that profoundly impacted him. Taking an ethics course was unusual for an engineer, but Cook wanted to broaden his horizons and develop a more global view of engineering and business. Even this early in his career, he was interested in the idea that companies could be a force for good in the world. This way of thinking was unique, as most science and technical education doesn't include any ethical training. The industry has always put an emphasis on technical skills, not social ones, and that has never been clearer than today, when the ethical nature of large tech companies' actions is under close scrutiny. From privacy abuses at Facebook to employee abuse at Uber, shoddy ethics in tech has become a hot-button issue. Starting around 2016, Silicon Valley has seen a tremendous backlash against the tech industry's ethos of moving at a lightning pace and breaking things with absolutely no regard for the consequences.

At a time when the integrity of the whole tech industry is being challenged on a profound level, Cook's ethical stance at Apple stands out. "When many people in business think of ethics, they think of accounting fraud, and they think of insider trading. . . . But this not what I think of," he said during a class reunion talk at Duke in 2013. "When I think of ethics, I think of leaving things better than you found them. And to me, that goes from everything from environmentally, to how you work with suppliers with labor questions, to the carbon footprint of your products, to the things you choose to support, to the way you treat your employees. . . . Your whole persona fits under that umbrella."

The ethics class caused him to think about business in a different way than most other people in the industry. The lessons he learned there—that you should leave things better than you found them, take care of the environment, and treat your employees with respect—underpinned his beliefs and would become hallmarks of his tenure as Apple CEO. At IBM, he was beginning to lay the groundwork for his leadership at Apple, which also included his interactions with his peers.

IBM Social Life

Over the years, Cook has often been described as someone who distanced himself from other people, but Daugherty and Addesso don't think that's true of his time at IBM. "I wouldn't characterize him as a loner," said Daugherty. "He's working with people, he's running a team." He was often in the middle of everything, socializing with his peers—especially his fellow recruits and HiPo employees. Addesso indicates that there was "a group of people that were all hired together that were very friendly. They went out together. . . . He had a couple of . . . real close lady friends when he was there, along with the guys. I mean they'd go out drink beer together and all. He's just an ordinary guy."

Cook was very sociable at IBM, and is remembered for his sense of humor. These days, his public persona is understandably more straitlaced and considered. His presentations and interviews tend to show his more serious side. He comes across as a no-nonsense, sober kind of guy, definitely not a joker. But in private, colleagues past and present say he has a humorous side and is quick to laugh at himself. Daugherty said, "He just was very thoughtful, levelheaded, hardworking but also . . . had a good sense of humor."

Ray Mays had a slightly different recollection. He said that at work, "people . . . enjoy[ed] working with him. He was smarter than anyone else, more aggressive in a positive way than anyone else, and worked harder than anyone else." All employees worked hard at the plant, but its social culture was very blue-collar. Mays said, "It was work all day, drink all night. That sort of socializing." But "Tim was not part of that. Can't really tell you much because I think he spent most of his awake hours working." Taking all accounts into consideration, it appears that Cook had a small group of friends whom he enjoyed socializing with after work, but he was more likely to be found working after hours than partying.

Despite having a close group of friends at IBM, Cook maintained his privacy when it came to his personal life. No one there knew he was gay. Addesso said he kept his sex life very private. In conservative North Carolina, and at an organization like IBM, with its strict dress code and processes, it wasn't typical for employees to be openly gay anyway.

"I never knew anything about . . . his sexuality," said Addesso. "A number of years after he left, a whole bunch of guys that used to work with him were [playing] golf . . . and [it came up in] conversation. . . . But nobody cared. Everybody liked him. He never pushed it. He was very private about it." He kept his private life private, and focused on his work. Though Cook has now come out publicly, he still leads a very private life, away from the spotlight.

Promotions at IBM

Over his dozen years at IBM, Cook was promoted several times and became a second-level manager. In the IBM hierarchy, shop floor employees reported to a first-level manager, and three or four managers would report to a second-level manager. Third-level managers would report to a plant manager, and plant managers would report to Daugherty, the general manager.

One Christmas, when Cook been there for eight or nine years, he ran the plant over the holidays. A lot of senior managers liked to take the week between Christmas and New Year's as vacation. But this particular year, the plant was under pressure to ship a lot of computers before the end of the year. Daugherty said, "We were scrambling. We had . . . to ship a large number of PCs to make our year-end quota. And when I left, we'd go on vacation for the week. I always put one person in charge. . . . And Tim volunteered and did just a fantastic job of managing the total shop in that week and hitting our production numbers. . . . That had a lot of pressure on him because we had to produce and we did. And he did."

The end-of-year rush was a "huge deal." Ray Mays said, "We had every available eighteen-wheeler on the eastern seaboard docking at our dock and loading up. Tim was there on the docks from noontime till after midnight making sure things got shipped. . . . And he handled it well. Typical cool, calm, and collected. Most people would go crazy and pull their hair out. Tim is even-tempered and [has a] good demeanor . . . given the pressures."

Senior management recognized Cook's work ethic. Like most of the HiPos, he ultimately ended up at IBM's HQ. Toward the end of his time at the company, he was promoted to director of fulfillment for North America.

Cook Moves to Intelligent Electronics

Cook worked for IBM for twelve years until October 10, 1994, when he accepted the role of chief operating officer of the computer reseller division at Intelligent Electronics in Denver. The company is long gone, but at the time it was a leading supplier of microcomputers, workstations, and other technology, worth $3.2 billion in 1995, the first full year Cook worked there. The now thirty-three-year-old Cook was responsible for the distribution of all of IE's products and services. It's not clear why he left IBM for a smaller, less famous company, but the prospect of being a big fish in a smaller pond and a big pay raise likely contributed. In his first year, 1995, he was given a base salary of $250,000, a $67,500 bonus, and one hundred thousand shares, a small fortune at the time, according to a 1996 filing with the Securities and Exchange Commission. "The move was so lucrative, Tim could not really turn it down," said Mays.

But his transition wasn't always so easy. During his time at IE, Cook suffered a major health scare. In 1996, he started to experience fatigue, tingling, and loss of coordination. Scarily, he was initially misdiagnosed with multiple sclerosis. "The doctor said, 'Mr. Cook, you've either had a stroke or you have MS,'" he recalled in an interview with the Auburn alumni magazine. The news made him "see the world in a different way," he says. It inspired him to take part in fund-raising events, such as an annual two-day cycling event across Georgia, to raise money for MS research.

It turns out that he didn't have MS at all; his symptoms were the result of "lugging a lot of incredibly heavy luggage around"—but he continues to raise money for MS research today.

After the misdiagnosis, Cook returned to his normal self and was able to enact major changes at IE. During his tenure, the company secured new distribution deals with Packard Bell, introduced a new pricing model, streamlined its operations, and launched a program called PowerCorps, which was specifically designed to increase sales of Apple products through IE's resellers. IE revenues had increased 21 percent to $3.2 billion in fiscal 1994 and climbed again to $3.6 billion in fiscal 1995. The company grew despite being hit by several lawsuits, one from a former employee alleging improprieties, and a class-action case from shareholders alleging that IE wrongly inflated its stock price by withholding information about its market practices. In 1997, Cook recommended selling the company to General Electric. It sold for $136 million.

Cook Joins Compaq

Shortly after the sale of IE, Cook was poached by Compaq, one of IE's suppliers, where he took up the position of vice president for corporate materials in Houston. By this point, Compaq had overtaken Apple and even IBM to become the world's biggest PC manufacturer, and it had just completed the acquisitions of Tandem Computers, known for its NonStop server lineup, and Microcom, a major modem vendor.

In February 1997, Compaq introduced the Presario 2000 Series, its first sub-$1,000 desktop aimed at the 60 percent of U.S. households that had never owned a PC. The entry-level machine ran Windows 95, and was somewhat of a gamble for Compaq, with the company choosing an unconventional but cheaper MediaGX processor from upstart Cyrix Corporation over more familiar chips from Intel. Intel chips were—and still are—the standard for most PCs. The Cyrix chip was from an unknown, no-name company, and therefore risky. But the gamble paid off. The Presario 2000 lineup would help Compaq, which was now employing more than thirty-two thousand people, reach $1.86 billion in profit for 1997.

The MediaGX processor helped kick-start the affordable computer trend, pushing prices under $1,000 and making computers more accessible for home users. It forced Intel to launch the Celeron, its own affordable central processing unit (CPU) brand, in April 1998, and AMD, another technology company, followed suit with cheaper chips of its own. PC prices fell steadily as a result, resulting in dire consequences for Apple. While PCs were getting cheaper, Apple's machines were not, and the company was in deep trouble, sitting on massive stockpiles of unsold computers.

During his short six-month tenure at Compaq, Cook helped the company transition to a build-to-order manufacturing model, another offspring of JIT, which Compaq named the "Optimized Distribution Model." Rather than building machines in anticipation of demand and allowing them to stack up on shelves, Compaq would begin the manufacturing process after orders had been received—a model that rivals like Dell and Gateway were already following. Instead of building to a forecast, computers were built to fulfill actual orders. This gave Compaq more flexibility and helped reduce waste, but it also meant the company had to manage its suppliers more effectively so that it could respond to demand quickly and cost-effectively. "We are sending a shockwave through the industry," said Eckhard Pfeiffer, Compaq's CEO at the time. "The new model will shape the way all Compaq products are designed, built, configured, distributed, ordered, purchased, serviced, and upgraded, as well as the way Compaq engages customers and works with its reseller partners."

Thanks to build-to-order manufacturing, which helped increase efficiency and dramatically lower production costs, Compaq was able to slash the prices of some of its most popular computers in the fall of 1997. Months later, in February 1998, it cut prices across its entire Deskpro line by up to 18 percent, forcing Dell, IBM, and others to follow suit soon after. "By continually finetuning our Optimized Distribution Model and garnering even greater efficiencies, we are able to deliver leading technology features and greater value across the Deskpro platform," said Michael Winkler, who was senior vice president and group general manager of PC products at Compaq at the time.

ODM allowed Compaq to hand off inventory costs to its manufacturing partners, which would ship complete machines only after they had been ordered. This negated the need for large warehouses, where computers had previously stacked up for lengthy periods while waiting to be sold. The savings Compaq made were better spent in other areas, explained Peter C. Y. Chow and Bates Gill in their book *Weathering the Storm.* "From Compaq's perspective, the adoption of the ODM enables it to concentrate on its own core competencies of R&D and marketing while leaving the rest of the value chain to its subcontractors in Taiwan and to vendors." It was essentially the model that Cook would later bring to Apple.

Cook played an important role in the ODM transition at Compaq, and it wasn't long before his efforts would land him on Jobs's radar. At Compaq, he was the middleman who had worked closely with the company's contract manufacturers to make the transition to ODM happen. At the time, Apple was in desperate need of overhauling its own messy manufacturing process. Jobs started looking for solutions and found just the right man for the job. "Tim Cook came out of procurement, which is just the right background for what we needed," Jobs later recalled to Walter Isaacson. "I realized that he and I saw things exactly the same way. I had visited a lot of just-in-time factories in Japan, and I'd built one for the Mac and at NeXT.

I knew what I wanted, and I met Tim, and he wanted the same thing." Cook and Jobs, two very different leaders, would come together over JIT manufacturing.

Cook kept to himself while at Compaq, but as at IBM, he was well liked. He lived alone in the city, while the vast majority of his colleagues had settled down with their families in Houston's suburbs. But before anyone really got a chance to get to know him, Cook left his safe and secure job at Compaq to join Apple, where he would apply everything he had learned to completely overhaul the way the company built and sold computers.

Chapter 4

A Once-in-a-Lifetime Opportunity to Join a Near-Bankrupt Company

When Cook joined Apple on March 11, 1998, it was not a place where many people wanted to work. The company was near bankruptcy, and employee morale was low.

Steve Jobs had recently rejoined the company as its interim CEO, or iCEO (he dropped the "interim" part of the title in 2000). His return to the company gave people a reason to be optimistic, but he hadn't actually sold anything yet. The only good thing to come out of Apple at the time was its celebrated "Think Different" ad campaign. Jobs was busy making changes within the company, stripping away the dead wood as Apple was hemorrhaging customers and revenue at a precipitous rate.

Apple's descent into trouble had been swift. Just four years before, in 1994, it was riding high. After IBM, it was the second-biggest company in the computer industry. It was raking in money from the desktop publishing revolution, selling boatloads of easy-to-use Macs to magazines, newspapers, book makers, and others in the publishing industry. Business was so good that Apple had three giant factories in California, Ireland, and

Singapore cranking out computers twenty-four hours a day, seven days a week for markets in the United States, Europe, and Asia. The company employed more than thirteen thousand people and had annual revenues in excess of $9 billion.

But on August 24, 1995, Microsoft released Windows 95, an operating system for IBM-compatible PCs that took the computing world by storm. Windows 95 was a blatant rip-off of Apple's Macintosh operating system, but it made cheap IBM clones from Dell, Compaq, and Gateway user-friendly. Windows 95 was a huge commercial success for Microsoft; in the first year alone, it sold forty million copies. PCs running Windows weren't as polished as Apple's machines, but they were a lot cheaper. They flew off shelves, and Apple's computers didn't.

The hit on Apple was almost immediate. Apple had made a profit of more than $400 million in 1995, but in the first quarter of 1996, it reported a loss of $69 million. This was followed by a much bigger loss in the second quarter of $700 million—one of the biggest losses in Silicon Valley history at the time. As a result, Apple started laying off workers, fired its luckless CEO, Michael Spindler, and hired a new CEO to replace him, Dr. Gilbert Amelio, who had a reputation as a turnaround artist, in the hope that he could save the company. But the slide continued. Over the next eighteen months, Apple's share of the computing market collapsed from 10 percent to an anemic 3 percent. Its stock nosedived. In the eighteen months Amelio was on the job, Apple lost $1.6 billion and was skirting with liquidation. But Amelio did one thing right—he brought Steve Jobs back to Apple, purchasing Jobs's company NeXT for $400 million to gain its next-generation operating system. At first, Jobs was brought in as a consultant to Amelio, but he shortly engineered Amelio's ouster and was asked by the board to rejoin the company he had cofounded years before as iCEO.

To save Apple, Jobs had to cut it back to the bone—compensating for the fact that Amelio had done exactly the opposite. In response to the huge selection of cheap Windows PCs on the market, Amelio had expanded Apple's product line to more than forty different models. There were four main product lines—the Power Mac, PowerBook, Quadra, and Performa—which in turn were split into dozens of different models. To customers, it wasn't immediately clear how each model differed from the others, and the baffling model names—Performa 5400CD, Performa 5400/160, Performa 5400/180 (DE), and so on—only added to the confusion. The company resorted to making elaborate flow charts to help customers decide. As Jobs pointed out on his return, "If I couldn't figure this out . . . how could our customers?" He had a point.

With Jobs back at the helm, unprofitable product lines like the Newton, an early handheld computer, and much of the engineering and marketing groups who worked on them were sent packing, though the best team members were reassigned to work on other projects. Out too went the disastrous licensing agreements for Mac OS, which had led to a spate of "clone Macs" in the years before.

Famously, Jobs also decided to cut the sprawling product line back to just four models: two desktop machines—one for ordinary consumers, the other for professionals—and two portable machines. He sketched his plan in a simple two-by-two matrix on a whiteboard. Apple's board of directors was spooked by the bare-bones plan. Jobs was putting all of Apple's eggs into just four products, when its competitors were shipping dozens. If any of the four computers failed, it could bring the whole company down. One board member said the plan was "suicidal."

Jobs's commitment to transforming Apple verged on mania. His neighbors in Palo Alto commented that on their nighttime jogs, they would rarely pass the Jobs residence without seeing Steve through the windows, staring at his computer screen, writing endless emails. By the time he came back to Apple, Jobs was a billionaire, thanks to Pixar's public offering a couple years earlier. But he also had a lot to prove. He was returning to the company that had kicked him out more than a decade earlier. And for much of the 1990s the tech press had gleefully focused on his failures, casting him as a "one hit wonder," particularly as his work at NeXT failed to achieve the commercial results he'd been hoping for. Jobs knew that there needed to be a major shift at Apple, and he worked tirelessly to achieve it.

Soon after he returned and was in place as CEO, a mischievous employee sent out a fake email from Jobs, parodying his views that Apple employees were "lazy and [contributing] to Apple's current situation." The email said that employees would now "have to pay for the water in our water fountains" and that a deduction would be made on paychecks for "the oxygen that you use for your eight hours on the job." Steve sent out a real email less than half an hour later. "I'm all for having fun," he wrote. "But we need to be focused on the future in making the company a better place. Best, Steve." The mischievous employee who sent the first email was fired.

Jobs was attempting to change the culture at Apple, one person at a time. But the way that he went about transforming the company was often perceived as arrogant. It appeared to some that he was pushing his own lifestyle on everyone who worked for him. Grueling work hours were expected, smoking was banned from the One Infinite Loop campus, and distinctly Steve-esque meals, many consisting of tofu, started to crop up regularly in the Apple cafeteria.

In addition to simplifying the products Apple offered and changing the culture within the company, Jobs knew that the key to its future success lay in overhauling the operations department, which ran manufacturing and had been suffering problem after problem in the preceding years. For years, the company had problems correctly forecasting what demand would likely be for its new computers. In 1993, it was burned by excess inventory for its PowerBook laptops, which were nowhere near as popular as Apple had forecasted. More disastrously, in 1995, it grossly underestimated demand for its next-gen Power Macs and was way too conservative in placing production orders. Forecasts were too low, there wasn't nearly enough flexibility in the supply chain to make up for it, and as a result, Apple was unable to provide enough machines to customers.

Preorders for the new 1995 Power Macs had been impressive, with 150,000 computers selling before a single one landed on anyone's desk. The first machine was a commercial and critical hit. In a four-star review, *Macworld* magazine described the Power Macintosh 6100/60 as such: "Not only has Apple finally regained the performance lead it lost about eight years ago when PCs appeared using Intel's 80386 CPU, but it has pushed far ahead." This was the same year that Windows 95 swept the PC world, so the thought that Macs were smashing the competition was welcome news for the struggling Apple. Sales were also impressive. In its first year of release, the Power Mac represented the top-selling multimedia personal computer on the market. In the twelve months starting March 1994, Apple sold 1.2 million Power Macs. The *San Francisco Chronicle* called it a "banner first year" for the new product line.

But it could have sold significantly more. Because of poor forecasting, Apple fell behind in production, and customers frequently had to wait up to two months to get hold of a machine. Business academics Robert B. Handfield and Ernest L. Nichols describe the problem in their book *Supply Chain Redesign:* "[Apple] was unable to obtain timely deliveries of critical parts, including modems and custom chips, and was not able to capitalize on the demand for its products." It was one of Apple's worst years in history—when it desperately needed the business, the company had $1 billion in unfilled orders in its system because it didn't have the necessary inventory capacity. The industry publication *Supply Chain Digest* ranked it as the one of the "greatest supply chain disasters" in history.

Back then, Apple insisted on custom-designing many of the components for its products, which were also outsourced from a single supplier. When things work well, custom-designed components can be a boon to a tech company, since high-performing custom components aren't available off the shelf to rivals and can't be copied easily. (Today Apple is investing more and more in custom parts, such as its own chip designs, for exactly that reason.) The downside of this, though, is that it leads to less flexibility. Fail to gauge the correct number of orders up front and the result is disastrous. Not only do you have to hurriedly assemble new products, but you also have to create the components that make up those products. As Handfield and Nichols observed, the Power Macs saw an increased demand of 25 percent, but managers only predicted 15 percent. That 10 percent difference alienated customers who were simply not willing to wait.

Apple's stock dropped precipitously as investors realized the company's mistake. Speaking with reporters in London, then chief executive Michael Spindler admitted that "we were a little timid" in forecasting sales of the Power Mac computers. As a result, he acknowledged, "We left some money on the table." His admittance was a gross understatement, and Apple's stock declined 8 percent to just $35 a share. By early 1996, Spindler was out as Apple CEO. Describing the debacle, the *San Francisco Chronicle* noted, "Investors just hate it when [this kind of thing] happens. There's nothing worse than having a hot product, and not being able to get it into customers' hands." Apple certainly learned this lesson the hard way. When Jobs returned to Apple in 1997, he was determined not to see the same mistakes happen again and started to look at transforming all areas of operations.

In 1997 Apple was running its own factories in Sacramento, California; Cork, Ireland; and Singapore. In theory, the idea was that all three factories would produce the same motherboards and assemble the same Apple products, which would then be sold in their respective markets (the United States, Europe, and Asia). But it didn't work quite as neatly as this. In some cases, products like the PowerBook were partly assembled in Singapore, shipped to Cork for more components, sent back to Singapore again for final assembly, and then sent on to the United States to be sold. It was a mess.

To save time and money, Apple started outsourcing manufacturing gradually, piece by piece, to countries like Korea and China. This was a seismic shift. Ever since the company's beginning, running its own manufacturing had been a core part of its identity. Jobs loved to have tight control over both hardware and software, and running Apple's own manufacturing was a key part of maintaining that control. The idea of outsourcing such a key part of the company was an anathema to him. Jobs had always been fascinated by factories, and had built two state-of-the-art just-in-time assembly plants during his career, with mixed success. As enthusiastic about end-to-end control as Jobs was, he was willing to compromise in the name of efficiency. First, in 1996, Apple sold its motherboard plant in Fountain, Colorado, to SCI and signed an outsourcing agreement that saw SCI supply circuit boards. A year later, Apple sold a circuit board factory in the Netherlands to NatSteel Electronics, a contract electronics manufacturer, and again entered an outsourcing agreement. In 1998, the year Cook joined Apple, PowerBook production was outsourced to Quanta, another contract electronics manufacturer, in Taiwan. This was just the start. Cook would accelerate the outsourcing process over the next few years.

Jobs's willingness to outsource and make the numbers work impressed a lot of the folks who remembered him as an impetuous youngster more concerned with "putting a dent in the universe" than making sure the lines in a spreadsheet added up. "He became a manager, which is different from being an executive or visionary, and that pleasantly surprised me," said Ed Woolard, the board chair who had helped bring Steve back to Apple.

Jobs's years as CEO at NeXT and Pixar had made him a far more efficient, practical manager than he had been during his first stint at Apple as a twenty-something. But he still needed help, particularly when it came to overseeing the kind of operational shift that would turn Apple into a winner again. That person, he concluded, could not be found among his existing lieutenants—a mixture of people he had brought with him from NeXT and a few who had been at Apple when he returned. But it took him awhile to find the right person. The first hire as head of operations quit after just a few months because of Jobs's abrasive style. Instead of hiring a successor immediately, though, Jobs began running operations himself to avoid taking on the kind of "old-wave manufacturing people" who applied for the post.

He wanted someone with the expertise of Michael Dell, the Dell Inc. CEO, who in 1997 had famously quipped that if he had been running Apple he'd "shut it down and give the money back to the shareholders." Jobs denigrated Dell publicly for his "rude" comment, but he admired him for his ability to build just-in-time factories and supply chains. But there weren't many candidates out there with Dell's expertise. It was at this point that Apple approached Tim Cook directly.

A Meeting of Minds: Cook Meets Jobs

Tim Cook had already turned down Apple recruiters multiple times, but their persistence paid off; he eventually decided that he should at least meet with Jobs. "Steve created the whole industry that I'm in," Cook revealed to Charlie Rose in 2014. "I'd love to meet him." Cook had been happy at Compaq, but his meeting with Jobs gave him a fresh and exciting outlook. Jobs "was doing something totally different," Cook recalled. As he sat and listened to Jobs's strategy and vision for Apple during their first meeting, he was persuaded that he could make a valuable contribution to Jobs's mission. Jobs described a product that he said would shake up the computing world, a design concept that would be unlike any computer seen before. The product turned out to be the enormously successful iMac G3, the bulbous, colorful Macintosh launched in 1998, which put designer Jony Ive on the map. Cook was intrigued. "He told me a little about the design, enough to get me really interested. He was describing what later would be called the iMac." Cook came away from the meeting convinced that working with a Silicon Valley legend like Jobs would be "a privilege of a lifetime."

Though he had some doubts at the back of his mind, they were not big enough to dissuade him from taking the job. "Any purely rational consideration of cost and benefits lined up in Compaq's favor, and the people who knew me best advised me to stay at Compaq," Cook revealed during his 2010 commencement address at Auburn University. "One CEO I consulted felt so strongly about it he told me I would be a fool to leave Compaq for Apple." In the interview with Charlie Rose, he again acknowledged that "there was literally no one around [me] that was inviting doing it."

But Jobs had already convinced him, and he knew that turning down a job at Apple would have meant turning down the opportunity to be a part of something special. "I'd always thought that following the herd was not a good thing, that it was a terrible thing to do," Cook recounted. "The way that he talked, and the way the chemistry was in the room, it was just he and I. And I could tell, I can work with him. I looked at the problems Apple had, and I thought, you know, I can make a contribution here. . . . So all of a sudden I thought, I'm doing it. . . . I was young at the time. . . . It didn't make sense. And yet, my gut said, go for it. And I listened to my gut."

Cook may have been analytically minded, but he was in awe of Steve's enthusiasm and aura from the start. "Five minutes into my initial interview with Steve, I wanted to throw caution and logic to the wind and join Apple," he said. "My intuition told me that joining Apple would be a once-in-a-lifetime opportunity to work with a creative genius." His intuition couldn't have been more right. At the 2010 commencement address at Auburn, he said, "Working at Apple was never in any plan that I'd outlined for myself but was without a doubt the best decision that I ever made."

Coming from a procurement background, Cook couldn't have been a better fit for Apple—and for Steve Jobs personally. Upon meeting Cook, Jobs understood immediately that they shared the same perspective on manufacturing. "[Cook] had the same vision I did, and we could interact at a high strategic level," he said. Jobs had found a partner that he trusted so much that he was able to "just forget about a lot of things unless [Cook] came and pinged me." It was a perfect match.

New Leader of Ops

Jobs hired Cook, aged thirty-seven, as senior vice president of worldwide operations in March 1998, with a base salary of $400,000 and a $500,000 signing bonus. Cook was given the sizable job of overhauling Apple's manufacturing and distribution. And it was one of the best hires Jobs ever made.

It was obvious from the outset that Cook was exceptional at operations, even before he was hired, said Greg Joswiak, the thirty-year Apple veteran. "I remember when Steve was interviewing Tim because he was coming back and telling us amazing things about operations that he was clearly learning from his interviews with Tim," Joswiak told me in an interview at Apple Park. "And so he was having literally an impact on us and some of the operational thinking that happened before he was hired."

Joswiak explained that Apple was "on the brink of bankruptcy" at the time, and that Cook was inheriting a mess. He admitted that operations was one of their worst areas: "We were terrible at it. Terrible at managing costs of it. Terrible at managing inventory. Terrible at managing billables." Looking back at the cockamamie system that he inherited, Cook said, "As you can imagine, the costs weren't so good, and the cycle times weren't that good." But that wasn't true for long. "In came this guy who knew a lot about operations and . . . in the typical Steve fashion, he'd hire the best," Joswiak said. "He had the ability to do it even when Apple was struggling." Joswiak also recognized that "Tim was super smart"—he wasn't just an operations guy who would "dot the I's, cross the T's" and "make the trains run on time" but someone who also had a "business mind," which Jobs demanded of all of Apple's leaders. "You have to have that business sense, and Tim had that." From the get-go, it appeared Cook was perfect for the role.

"To this day I remember meeting Tim," said Deirdre O'Brien, another thirty-year Apple veteran, who at the time of Cook's joining was heading up Apple's supply and demand team. "And right off the bat it was very clear that he was very focused. He was incredibly excited about being at Apple. . . . He had a big job to do. You could tell he knew he had a mission."

O'Brien was hired by Apple in the late 1980s—four years after the original Macintosh launched—to do production planning in the Fremont factory, which was producing the Macintosh SE at the time. Over her long career, she has risen through the ranks in operations, working for Cook and then his successor as COO, Jeff Williams. She has worked under five different CEOs—John Sculley, Michael Schindler, Gil Amelio, Jobs, and Cook. She is now Apple's head of HR, the so-called Veep of Peeps—a title she said was Cook's idea.

O'Brien said she first met Cook at a get-to-know-you session in a conference room with a handful of Apple's operations staff. The meeting was brief; Cook introduced himself and let the staff know just how much of a challenge lay ahead. He said that there would have to be a lot of changes, and that included cuts to the number of staff.

But O'Brien wasn't disheartened by this potentially negative stance. She took the opposite message, as it was obvious to her that Cook's changes would get Apple back on track and ultimately lead to growth, not downsizing. "I don't think anyone would've taken that job if they did not think it was a huge turnaround role, and I think for Tim that was probably what was really exciting about it," she said in an interview at Apple's new spaceship HQ. "We were all so happy that we had a leader of operations. We had a lot of changes in the leadership . . . so it had been quite chaotic actually. . . . And at that time it was tough to attract great talent to Apple, and we had heard good things about Tim. . . . It was very clear to me right from the start that Tim was going to be someone that you could learn so much from."

After the initial get-to-know-you meeting, Cook met with individual members of the team to assess their strengths and weaknesses. "He was in assessment mode," said O'Brien. He was figuring out, "What do I have here? And how am I going to find a way to make sure that we can be successful?"

At the time, she said, a lot of people were leaving Apple voluntarily. There were plenty of jobs in Silicon Valley, and a lot of companies were happy to pick up ex-Apple folks. She said of the numerous departures that some people were "opting out and other people . . . Tim felt were not the right fit for the team he ended up assembling."

He worked hard to pull together the right team that would fix Apple's operations problem. O'Brien became part of the core operations team, running demand forecasting, and she was joined by several IBM alums, including Jeff Williams, who became Cook's right-hand man and is now

Apple's COO; Bill Frederick, a logistics expert, who headed up customer support; and Sabih Khan, who ran notebook operations. It was only a matter of time before this handpicked team began to turn things around.

Goodbye, U.S. Manufacturing. Hello, China!

Just seven months after arriving at Apple, Cook had reduced inventory from a whopping thirty days' worth to just six. In a short amount of time, he performed a complete overhaul of Apple's operations system by paying attention to every single detail of the production process.

Jobs's decision to cut Apple's product line down to just four models made the company's operations problems much simpler. The computers laid out in Jobs's two-by-two product matrix purposely shared many components and used industry-standard parts where possible. Previously, Apple had been notorious for inventing its own esoteric technology that was incompatible with other systems. Ten years later, gone were the Apple-only ADB keyboards and mice, replaced instead by industry- and Windows-friendly USB devices. Not only did this help from an operations perspective, but it also had the added benefit of making Apple products much more compatible with the wider computer market.

Cook used the same focus that Jobs had used to whittle down Apple's product offerings to single out only a few suppliers to work with. He visited each and every supplier, and though he struck hard bargains, he also charmed them. He persuaded NatSteel, Apple's outsourced circuit board supplier, to set up plants close to Apple's own factories in Ireland, California, and Singapore. Moving suppliers closer to the plants made the just-in-time process much easier, because components could be delivered faster and more frequently.

Wherever possible, he outsourced. The iMac G3, for instance, was initially built at Apple's own factories, with the exception of the case and monitor, which came from LG Electronics. When Cook took over, he outsourced most of the production to LG. He also outsourced laptop production to two companies in Taiwan. Now Quanta Computer would produce the pro-level PowerBook, and Alpha Top Corporation would produce the consumer-oriented iBook.

Outsourcing production to outside partners helped Cook solve one of Apple's biggest problems: inventory. The company had parts in storage at one end and unsold computers at the other. Warehouses of parts and unsold machines had cost the company millions in inventory costs over the years. Big stockpiles of unsold computers had nearly bankrupted Apple in 1996, so under the new regime, Cook thought, the less inventory, the better. "Warehouses tend to collect product," he said. "We started to deliver directly from manufacturing to the customer."

Cook hated excess inventory with the same zeal that Jobs hated shoddy design. He even cast it in moral terms, describing piled up inventory as "fundamentally evil" because of its drag on company finances. He gave an analogy that seemed to have been inspired by his upbringing in rural farm country. "You want to manage it like you're in the dairy business," he said. "If it gets past its freshness date, you have a problem."

Good inventory management is underpinned by the ability to forecast sales in advance, but Apple had a mixed history in forecasting. The company had a long, sad history of making too many or too few computers at various times. For example, Steve Jobs's original Macintosh factory in Fremont, California, dubbed the "Factory of the Future," claimed to be the best in the computer industry. But it was shuttered after a few years because Apple never sold enough computers to meet the factory's capacity.

To get a handle on forecasting, Cook invested in a state-of-the-art enterprise resource planning (ERP) system from SAP that hooked directly into the IT systems at Apple's parts suppliers, assembly plants, and resellers. The complex system gave his operations team an extremely detailed bird's-eye view of the entire supply chain, from raw materials to customer orders at Apple's new online store, which had recently launched. The R/3 ERP was the central nervous system of Apple's new, lean, just-in-time production. Parts were ordered from suppliers only when they were needed, and factories produced only enough machines to fulfill immediate demand.

Armed with mountains of data, Apple's operations team started tweaking production on a daily basis, sending orders to contractors based on detailed weekly sales forecasts and extremely accurate tallies of stock in retail channels. Cook's team could tell if CompUSA was nearly running out of iBooks or sitting on a mountain of iMacs.

Under Cook's leadership, the amount of time that Apple's inventory sat on the company's balance sheet was reduced from months to mere days. In the seven months after he started work at Apple, thanks to Cook's achievements slashing inventory turnover from thirty days to six, the company's inventory stock was reduced from $400 million worth of unsold Macs down to just $78 million. In 1998, Cook got rid of tens of thousands of unsold Macs, which had carelessly piled up in the days before Apple's turnaround, by sending them to a landfill. The episode is shrouded in mystery, and understandably, Apple kept it very quiet. It's at odds with the environmentally friendly principles that Cook has instilled at Apple today. But it was very effective at the time. By 1999, inventory had been reduced to just two days' worth, and Apple was beating Dell in this department—an astonishing feat, given that Dell was considered the industry gold standard.

Now that operations had vastly improved, Cook was credited with playing a key role in Apple's return to profitability. His system also fueled Apple's phenomenal growth in the years after. It's clear that Apple would never have grown so big and dominant without Cook's excellence in operations. Just as Jony Ive's industrial design team crafted great products, Cook's team figured out how to produce them in vast quantities—and have them delivered to stores all around the world, without delay and in the tightest secrecy. Apple is famous for keeping its unannounced products under wraps until just before they go on sale. It's no mean feat that Cook's operations team is able to maintain its secrecy while millions of products are being quietly made and shipped to stores worldwide.

But Cook contributed more to Apple than just streamlining the production process to be more efficient—he innovated at every twist and turn. One of his earliest and biggest coups—and a sure sign that he was "thinking different" within his own department—was a neat bit of jujitsu involving shipping of the iMac G3. Apple was trying to make the iMac into a mainstream product, and in order to do that the company would need to get it onto as many people's desks as quickly as possible. To make sure that the computers shipped out to customers in an expedient manner over the all-important holiday season, Cook booked $100 million worth of air freight months in advance. This was unheard of, but it paid off in a big way. Not only did Apple get its products out to customers in a rapid fashion, but its rival PC makers, such as Compaq, suddenly found themselves struggling to secure shipping over the holidays. Thanks to Apple's new approach to booking shipping space, other companies were forced to rethink their own operations strategy. Cook not only improved Apple's operations, but he also changed the way the entire technology industry managed its production process and the way that process was perceived in the industry.

Back when Cook joined Apple, forecasting demand and improving supply chains wasn't cool in the same way that creating candy-colored computers was cool. No one was lining up to make Apple's operational overhaul the cover story on *Fortune* or *Wired*, and to the average customer, the supply chain side of a business is only noticeable when it fails to work as promised. But against all odds Cook made operations cool. Gautam Baksi, a former lead product design engineer, said that in the very early days of Apple in the buildup to the iMac, when the teams would visit China, "all the designers, including Ive and Danny Coster, lived and ate in the same hotels as all the engineers and the rest of the Apple crew." But ten or fifteen years later, when Cook joined Apple, "the ID guys got picked up in stretch limos" and stayed in "five-star hotels," while the engineers "had to take a taxi" and stayed in "three-star hotels." It was clear that before Cook, operations was seen as a less important, less glamorous part of the business. But that all changed in the Cook era. Now it's the operations people who are riding in the stretch limos and staying in five-star hotels.

Chapter 5

Saving Apple Through Outsourcing

A pple changed dramatically during Cook's first year at the company. After reporting a net loss of just over $1 billion in 1997, it was already seeing profits by the end of 1998. The runaway success of the first iMac exceeded Apple's own expectations, and those of Wall Street analysts, pushing Apple to make a profit of $106 million for the fourth quarter, and $309 million for the fiscal year. "Apple grew faster than the industry this quarter for the first time in nearly five years," Jobs boasted. He attributed that to the new iMac and the company's new streamlined business focus.

But despite the profits, the company was still on rocky ground and was trying to make savings everywhere it could. Cook reviewed Apple's capabilities product by product and started outsourcing more and more. The factories were expensive to run and a liability on the balance sheet. It made sense to offload as much as possible to outside suppliers, without sacrificing quality or productivity.

Deirdre O'Brien, Apple's VP of People, who at the time was in charge of the supply and demand management, said that Cook and the operations team "worked very hard to build something that . . . really supports our customers and builds incredible products." They did a thorough review of the company's internal strengths and the strengths of the suppliers they were working with. Cook and his supply team didn't just outsource Apple's manufacturing wholesale to third-party suppliers; they pursued a unique hybrid model. "It was not a traditional outsourcing model where you just hire someone and sort of toss the plan over to them, or, in many cases, you actually take their plan and put your name on it," she said.

At first, iMac production was partially outsourced to LG, which made the computer's cathode-ray-tube screen and other components. In 1999, LG took over iMac production completely, and as orders and demand increased, Apple brought on another Taiwan-based contract manufacturer best known at the time for working with Apple's rival, Dell. Hon Hai Precision Industry Company Ltd., better known as Foxconn, would come to define manufacturing in the Tim Cook era. Although Apple had worked with Foxconn years before, on assembly of the Apple II, the iMac contract marked the beginning of the two companies' transformative relationship, which Cook spearheaded.

Foxconn

Foxconn was founded around the same time as Apple, some six thousand miles away on the other side of the world. In 1974, when nineteen-year-old Steve Jobs was working at Atari, twenty-four-year-old Terry Gou borrowed $7,500 ($37,000 in today's money) from his mother to start up a business.

Gou has a work ethic second to none, and he expects the same level of commitment from his employees, so the culture at Foxconn is militaristic. Orders are expected to be followed to the letter, and there is no tolerance for mistakes or inefficiency. If a worker makes a mistake, he would be reprimanded publicly in front of the other workers. If the worker makes the same mistake twice, he would be fired. The hours are long and punishing; shifts are typically twelve to fourteen hours, and workers often toil six days a week, sometimes seven.

Most consumer electronics, including all of Apple's products, are almost entirely assembled by hand, and it's a misconception that a lot of machines or robots are involved. Some parts are made by highly automated processes—the iPhone's main logicboard, for example—but most of the assembly, especially the final assembly, is done by hand. Anna-Katrina Shedletsky, a former Apple product design engineer, said, "Hundreds of hands touch a smartphone, regardless if it is an Apple smartphone or Samsung or Google smartphone. This is state of the art for assembly of these complicated, highly miniaturized devices."

At Foxconn and other assembly plants, the standard assembly line is 110 meters long—a little bit longer than a football field, according to Shedletsky. It is divided into sixty or seventy stations, each about 0.6 meter wide—the standard width of a human being. Each station is devoted to a specific step or operation in the assembly process. The worker at one station snaps the screen into place; the worker at the next station cleans the newly installed screen with a solvent to remove dust and oils; a third worker applies a protective film; and so on. There are twenty-foot gaps on either side of the line, wide enough to allow forklifts with pallets of parts to access each station and restock components.

To many in the West, Foxconn's factories are almost unimaginable in scale. They are enormous complexes, complete with sleeping quarters, restaurants, hospitals, supermarkets, and swimming pools packed into, in the case of Foxconn's Shenzhen factory, a 2.3-square-kilometer space. They are more like factory towns, or, as CNN once described it, a "heavily secure" university campus. Duane O'Very, a Foxconn manager in the late

1990s and early 2000s, said he saw the campus explode in just a few years from about 45,000 workers to more than 250,000. Foxconn now employs an estimated 1.3 million in China alone, where it operates twelve factories. It also has plants elsewhere in Asia, and in South America and Europe.

It's often assumed that Apple and other companies assemble their products in China because of low labor costs. The cost of labor is a fraction of the cost of the materials that make up the iPhone. Far more is spent on the custom chips, intricate cameras, and beautiful screens than on the labor required to assemble them.

But the key to Foxconn's success isn't cheap labor. It's flexibility. Because Foxconn's compounds have hundreds of thousands of workers living on site, the company can be very flexible, assembling armies of workers literally overnight. Foxconn is also able to hire tens of thousands of extra workers very quickly, and fire them just as quickly if they are not needed. Its young workforce is often recruited from far-flung rural districts, so it's not always easy for them to quit and go home, even if the work is soul crushing.

Foxconn demonstrated this flexibility during production of the first iPhone, which underwent a major design change just a few weeks before it was scheduled to go on sale in 2007. At the last minute, Steve Jobs decided that the iPhone should have a glass screen instead of a plastic one. He'd been carrying a prototype iPhone in his pocket for weeks, and its plastic screen had been badly scratched by his keys. He knew this would be a problem when the phone got into customers' hands, so he demanded a more durable glass screen.

A few weeks later, the new glass screens arrived at Foxconn in the middle of the night. More than eight thousand workers were roused from their beds, given a cup of tea and a biscuit, and started on a twelve-hour shift fitting the new screens to phones, according to the *New York Times*.

Within a few days, the factory was outputting more than ten thousand iPhones a day. (It's worth noting, though, that Foxconn has denied that this event ever happened. They claimed that Chinese labor law would have made the episode impossible.)

On another occasion, when Foxconn was building the first iMac for Apple, the company's design engineers were scrambling late at night with a new button for the machine. The button was untested, and the designers were worried it might fail with continued use. So about a dozen Foxconn workers were wakened and instructed to repeatedly press the button all night long to test it. "That's very easy for [Foxconn] to do," said former product design engineer Gautam Baksi. "At other factories, they might want to design a machine that constantly pushes it, but it's no problem asking a minimum-wage worker to push the button all night long. That happens routinely. That's not even a stretch example of what they'll do."

Apple often makes last-minute design changes to its products, and demand can fluctuate wildly, especially when the numbers are so huge. The iPhone X sold an estimated fifty-five million units in its first few months. At that scale, Apple needed factories capable of producing up to a million units a day at peak volume, which would require an estimated 750,000 workers. "They could hire 3,000 people overnight," said Jennifer Rigoni, a former Apple worldwide supply demand manager. "What U.S. plant can find 3,000 people overnight and convince them to live in dorms?"

Apple certainly couldn't. In the 1980s, Apple's factories were highly automated, and Steve Jobs learned this lesson the hard way. Back then, Jobs set up an automated (and highly publicized) factory in the Bay Area to make the first Macintosh. It was a gleaming jewel, with color-coordinated machines and a highly automated line.

The production line was built from automatic conveyor belts and lots of exotic machines for picking, packing, and moving Macs around. But unfortunately, sales of the Mac never picked up enough to cover the costs of the factory. Sales were so anemic that the factory operated well under capacity, and because it was so specialized—designed to make one machine, and one machine only—it couldn't be reconfigured to manufacture other products. This was what ultimately doomed the factory, which closed in 1992. Jobs's highly automated factory would not have been flexible enough to manufacture different products, such as the iPhone, iPad, and iMac, according to fluctuating demand. Cook's revolutionary outsourcing initiatives decreased the need for domestic factories and contributed hugely to Apple's increased success. By outsourcing the majority of Apple's operations and ramping up their partnership with Foxconn, Cook was doing something that had never been done before, to amazing results, and Apple executives, Steve Jobs especially, took notice.

Cook Climbs the Ladder

Cook's transformation of Apple's operations and deep understanding of every aspect of the business was pivotal to the success of the company's dramatic comeback. His experience heading up this instrumental department of Apple would prepare him to lead the entire company later as COO and then CEO. His leadership potential was immediately apparent, said Deirdre O'Brien. "Now it seems very clear that he could become our CEO," she said during an interview at Apple Park. "At the time he came in . . . he . . . wasn't only worried about how to optimize operations. He cared about everything." He coordinated with the engineering team and the sales operation to understand both the products and the customers who purchased them.

In his first couple of years at Apple, he helped oversee the launch of memorable Apple products, including the bulbous iMac G3, the blue-and-white Power Macintosh G3, the toilet-seat iBook laptop, and more. Each turned out to be a hit with customers and critics alike. In September 1999, just two years after Jobs had returned, AAPL stock hit what was then an all-time high of $73 a share, beating the previous record of $68 set in 1991. The spectacular turnaround of Apple, spurred on by Tim Cook, was under way.

In 2002, four years after Cook joined Apple, he was tasked with leading sales as well as operations. He was also given a new title: executive vice president of worldwide sales and operations. In 2004, Jobs appointed him head of the Macintosh hardware division, and in 2005 he secured another big promotion to COO. "Tim has been doing this job for over two years now, and it's high time we officially recognized it with this promotion," Jobs said at the time. "Tim and I have worked together for over seven years now, and I am looking forward to working even more closely with him to help Apple reach some exciting goals during the coming years."

With the promotions, Steve Jobs was grooming Cook as his successor. Everyone at Apple is a specialist, with two exceptions: Steve Jobs and Tim Cook. Apple is a functional organization, and all the staff specialize in narrow disciplines: cryptographic programming, industrial design, antenna engineering. Cook was initially an operations specialist. The only person in the organization without a specialty was Steve Jobs, until he pulled Cook out of operations and gave him more responsibilities. With Cook's promotion in 2002 to VP of sales, and then to head of the biggest hardware division at the time (the Mac division), Jobs was training him in all the different parts of the business. Cook's successive promotions were a CEO apprenticeship, culminating in his appointment to COO in 2005, when he officially became Jobs's right-hand man.

Cook took on a lot more responsibility than the typical corporate COO. There's no single operations department at Apple—the term is an umbrella for a bunch of different groups that handle manufacturing, distribution, and service. The biggest group within that umbrella is the supply chain team, a cross-functional organization that is responsible for managing Apple's vast contract manufacturing operation.

Within this group are lots of smaller groups responsible for different aspects of the production process. One is a design manufacturing team, which makes sure proposed products can be manufactured at scale. It includes manufacturing engineers, process engineers, and quality engineers, among others. There's also a yield team, which is responsible for maintaining the quality of products or components coming off assembly lines. To ensure that supply meets demand, there's a planning department that forecasts projected sales and helps figure out—in extreme detail—all the resources needed to meet its projections, from the tonnage of recycled paper for packaging to the number of camera modules for a big iPhone launch.

Operations is perhaps the largest of Apple's divisions, but it's difficult to estimate the size of the group, and Apple doesn't publish an org chart—internally or externally. One ex–operations staffer estimated that operations might be as big as thirty or forty thousand people, the vast majority of the fifty thousand workers that Apple has based in and around Cupertino. As COO, with as many as forty thousand workers under his management, Cook would have had a huge influence over the entire company's culture.

Cook the Manager

Though Jobs and Cook worked together closely for many years, they were very different in manner and temperament, especially when it came to their management roles. Cook's approach was different from Jobs's, but it garnered results. Steve Jobs was the kind of person who would (and did) call Apple chip manufacturers "fucking dickless assholes" to their faces if they failed to deliver sufficient quantities of chips on time. He would rant and rave at people, belittling and insulting them as "shitheads" who could do no right.

Cook's tactics were markedly different. He rarely raised his voice, but he was relentless in attacking a problem—and could wear down people through an endless barrage of questions. "He's a very quiet leader," said Joswiak. "Not a screamer, not a yeller. . . . He's just very calm, steady, but will slice you up with the questions. You better know your stuff." The questions helped Cook really drill down into the issue and ensure that the staffers knew what they were doing. It was effective because it kept staff on their toes and on top of their responsibilities. They knew they might be called on to explain at any time. "He'll ask you ten questions," said Steve Doil, who joined Cook's operations group in December 1998. "If you answer them right, he'll ask you ten more. If you do this for a year, he'll start asking you nine questions. Get one wrong, and he'll ask you 20 and then 30."

Cook required incredible amounts of detail from his staff. "They're nervous going into [meetings with him]," said a person with knowledge of Cook's group from this time. "He'll say, 'What's this variance on column D, line 514? What's the root cause of that?' And if someone doesn't know the details, they get flayed right there in the meeting." A manager from Apple's hardware group was once shocked to listen in on a meeting headed up by Cook. At one point, an underling came up with a figure Cook deemed was incorrect. "That number is wrong," Cook said. "Get out of here."

As COO, Cook expected his teams to work hard, be proactive, and pay attention to every detail. And managers followed his lead, adopting many of his leadership techniques and expecting the same of their employees. Helen Wang, a former global supply manager, said that, like Cook, a lot of senior operations managers were detail-oriented, with an uncanny knack for numbers. Wang often saw senior leadership memorize entire spreadsheets or zero in on obscure cells with an aberrant number. They had the uncanny ability to spot a problem in a sea of numbers that could easily have been overlooked. Managers would often remember numbers from meeting to meeting and would question the supply managers if one changed. She said that managers would also adopt Cook's habit of asking question after question. "They want to know [if] you understand the problem," and "if you're not detail-oriented, I don't think you can even survive in that company." Cook "definitely created a lot of process that practices how you are thinking about a problem as well as the culture and the norms of the company," she said. "A lot of time you hear people say, 'That's how we do it.' I think that's either inspired by him or impacted by him. The way we think and how we do things."

Despite this emphasis on the importance of attention to detail and solving problems, Cook trusted and empowered his staff to make decisions for themselves. He inspired a mentality where, according to Wang, "everything is possible, let's try harder, let's be creative, let's try to solve it. You know we can do it. There's this can-do attitude. . . . People at the leadership level constantly reminded us to be creative. How do you solve the problem?" Wang said that although she was young (in her early thirties at the time), she felt grateful that Apple's management trusted her—and other staffers—to tackle issues without being micromanaged.

"The senior director level empowers their people," she said. "You always feel like, even if you're young, you might be junior . . . that whatever the role you are, you feel like you are making decisions in the best interest of the company, and the company trusts you. This is how Tim and Jeff [Williams, Cook's VP of operations] lead, how much trust they put on you."

To win Cook's respect and appreciation, employees not only needed the right answer to questions at all times, they also needed to show a willingness to go the extra mile. Sometimes literally. One classic illustration of Cook's firm approach came during a supply meeting when an issue arose concerning a manufacturer in China. "This is really bad," Cook said. "Someone should be in China driving this." The meeting continued for another half hour before Cook looked directly at Sabih Khan, a key operations executive, and asked, with deadly seriousness, "Why are you still here?" Khan immediately got up, left the meeting, drove to the airport, and booked a flight to China with no return date. He didn't even stop at his home to pack a change of clothes.

Unlike some of his colleagues, Cook appeared to have little life outside of Apple. He led by example, but it was an example that few—particularly those with kids and partners at home—could hope to follow. He took conference calls on Sunday nights, was replying to emails by 3:45 a.m., and was at his desk by 6 a.m. every morning. He worked twelve- or thirteen-hour days in the office, and then returned home to answer more emails.

"I would get a couple of emails from Tim between about 3:45 and 4:15 in the morning," and "then from 4:30 to 6:00 it would go quiet," said his colleague Bruce Sewell, Apple's former general counsel. "That's when he's at home and eating breakfast, getting up, getting ready to go to the gym." Then from about 6:15 onward he would be at work.

It wasn't unheard of for Cook to fly to China, work three days without acknowledging the sixteen-hour time difference, fly back, land at 7 a.m., and be in the office for a meeting at 8:30. When he wasn't flying to China to meet with Apple suppliers, he rarely left the state of California so that he could be available at a moment's notice. To him, meetings were more like marathons, an analogy made more appropriate by the fact that he snacked on energy bars throughout them. When he wasn't in the office, his idea of relaxation was either hitting the gym or going rock climbing. He's also an avid cyclist and often bikes on Saturdays and Sundays, providing colleagues a rare respite from his emails. "He exercises a lot," noted Sewell. "Tim's a very, very health-conscious person, so he gets up early because he wants to get to the gym before it gets crowded and before everybody else is getting up."

It's fair to say that Cook has treated Apple like a sport. Work was (and remains) a form of endurance sport to him—and it was evident in everything he did, even down to the way he clipped his hair short, reminiscent of one of his then sporting heroes, Lance Armstrong. In one operations meeting, he put up a slide with a quote from Armstrong: "I don't like to lose. I just despise it." He also equated work and sports in the Auburn commencement speech he delivered in 2010, the year before he became Apple's CEO. "In business, as in sports, the vast majority of victories are determined before the beginning of the game. We rarely control the timing of opportunities, but we can control our preparation." Cook's obsession with preparation was key to his success at Apple.

The first twelve years of Cook's career at Apple were relatively quiet. Jobs had always been front and center of Apple's publicity, and other key figures like Jony Ive had their own public image too, but Cook's appreciation for anonymity meant he kept himself hidden behind Apple's secretive curtain. That didn't change much in January 2009, when he first assumed the role of interim CEO after Jobs was forced to take a six-month medical leave of absence following a liver transplant. Jobs remained CEO and assured Apple employees in an internal memo that he would still be involved in "major strategic decisions," but it was up to Cook to take responsibility for Apple's day-to-day operations. "I know he and the rest of the executive team will do a great job," Jobs wrote.

During Jobs's absence, Cook oversaw the launch of the iPhone 3GS, which quickly became Apple's fastest-selling smartphone up to that point, with over one million units sold over launch weekend alone. Jobs returned to Apple in time to host the September 2009 keynote, during which he thanked the company's executive team, and in particular Cook, who "rose to the occasion and ran the company very ably." Cook did such a good job, in fact, that he was again temporarily placed in charge of Apple when Jobs took another medical leave of absence in January 2011.

Even when Jobs was sick and Cook acted as CEO, he stayed out of the public gaze as, ultimately, Jobs was still able to act as the face of the company. All this changed when Cook took over the top job.

Chapter 6

Stepping into Steve Jobs's Shoes

ook's first day as Apple CEO was Wednesday, August 24, 2011. Some of Cook's first moves as CEO, little noticed at the time, signaled the big changes that were to come. They would both mark him out from his predecessor and give rise to the astronomical success we associate with Apple today.

But though Cook was an expert at running Apple behind the scenes, he didn't have much experience being in the spotlight. The first year of his tenure as CEO was rocky, marked by wooden public appearances, executive shake-ups, two high-profile firings, and unexciting products. It didn't augur well with critics, who continued to predict that Apple was in a long, slow decline.

Wooden as Pinocchio

The first few months of 2012 dealt Cook more than a handful of headaches.

Headlining his first Apple keynote since Jobs's death, he introduced the iPad 3 and an updated Apple TV in March. "I'm very excited to be here," he said as he came onstage, but he looked anything but. Dressed in a rumpled, untucked shirt, he methodically plowed through his presentation in his unassuming southern drawl. Showing none of the charisma or magnetism that had made Steve Jobs's presentations so dynamic, he appeared uncomfortable and overrehearsed. He wasn't having much fun. His brow was furrowed and his delivery serious. Somehow, he managed the impossible: to strip an Apple keynote of its usual excitement.

The new iPad had great specs (a mix of new and updated features, including a new, high-resolution Retina display, a new Apple A5X chip with a quad-core graphics processor, an upgraded 5-megapixel camera, HD 1080p video recording, voice dictation, and support for LTE networks in the United States), but despite positive reviews, fans thought it was an underwhelming update that too closely resembled its predecessor—that it was more evolutionary than revolutionary. This was not good for the first major new product launched under Cook's leadership.

Early Setbacks

The first few months as a new CEO at any company must be a challenging time for anyone, let alone when the company's visionary founder has tragically died and the company is one of the most visible in the world. That Apple was sued by the U.S. Department of Justice made Cook's first few months as CEO even harder. In April, the DOJ accused Apple of conspiring with a number of book publishers in e-book price fixing. The case continued for several years and resulted in Apple paying a fine and taking on a court-appointed antitrust monitor. It also highlighted how Apple was now big enough to attract the attention of antitrust regulators. Antitrust cases are usually reserved only for the biggest, most powerful companies that are deemed to be abusing their position of dominance and need to be reined in. An antitrust case, settled in 2001, had unseated the once-dominant Microsoft from its position atop the tech heap, so this was considered a serious concern.

Then, in July, Apple shares dived after a disappointing Q3 redundant report that saw less-than-predicted iPhone sales. Analysts had expected Apple to sell 28.9 million iPhones, and when only 26 million sales were reported, shares dropped. It wasn't a big deal—iPhone sales were up nearly 30 percent from the year before—but it was only the second time in almost a decade that Apple had missed Wall Street's expectations.

Apple's disappointing iPhone sales may have been due to growing competition from Android. While Apple was still on top, Samsung was quickly becoming a threat, with the company's name surfacing more and more in articles about Apple. In May 2012, Apple was named the world's most valuable brand by Millward Brown's BrandZ study for the second year in a row. "Apple continues to innovate and maintain its 'luxury' brand status, but faces future competition from Samsung," the study said. "Now worth more than $14.1 billion, thanks in part to the success of its Galaxy handsets, Samsung is successfully outpacing Apple in a significant number of markets by positioning as a cool, well-priced alternative to the ubiquitous iPhone." In October 2012, Cook fired the head of Apple's Korean division, Dominique Oh, due to sluggish sales in South Korea, home of Samsung. Oh had been in the job a mere seventeen months.

Apple's declining share prices may also have resulted from missteps with its new products. In July 2012, Apple finally pulled the plug on MobileMe, its ill-fated cloud service, which had been a bust from virtually day one. Though it had already been replaced by iCloud in October 2011, MobileMe remained active until the middle of 2012. At the end of September, Apple officially shuttered Ping, the music-focused social network it had launched two years earlier as part of iTunes 10. Ping encouraged users to follow artists and friends to see what was popular and get music recommendations. But from the get-go it was marred by problems. Integration with Facebook was promised but never delivered. Some users' accounts were flooded with spam, and scammers started opening fake accounts in other people's names. But the biggest problem is that it failed to take off; only a fraction of iTunes users used it. Speaking earlier in the year at the All Things Digital Conference, Cook admitted that Apple didn't "need to have" a social network on its books. "Some customers love it, but there's not a huge number that do, so will we kill it?" he asked. And shortly after that, Apple killed it.

Hiring and Firing

The year 2012 also saw two high-profile executive firings. The first of these was of Apple Store senior vice president of retail, John Browett. He had only officially joined Apple in January, replacing previous Apple Store retail boss Ron Johnson. On paper, Browett had the right qualifications for the job. He came from Dixons Retail, one of the largest consumer retailers in Europe, with forty thousand employees. He had a degree from Cambridge University and an MBA from the Wharton School. An MBA who was good with numbers, he shared a lot of similarities with Cook in an earlier job. When he was working for British supermarket Tesco's online operation, he told a reporter, "I don't do lunch. I don't do conferences [because] there is too much to do." He had a no-nonsense approach to work that could have come right out of Cook's playbook. He seemed like a perfect fit to run the Apple Store under Cook.

By 2012, the Apple Store was more than a decade old. Though it had been met with skepticism when it launched in 2001, there were now four hundred locations around the world, with one-third of them outside the

United States. Sales figures per square foot put Apple Stores above any other U.S. retailer, including the luxury brand Tiffany & Company. The Apple Store had changed the way that computers and other electronics products were sold, and had given Apple the end-to-end control over the retail experience that its executives had always wanted.

The store was a hugely important part of Apple's business, and Browett was Cook's first high-profile hire as CEO. Almost immediately, however, he ran into problems. It seemed from the start that he wasn't a good fit for Apple's culture. Within a day of announcing his hiring, Cook began receiving worried emails from customers, concerned that Apple Store quality would decrease with Browett at the helm because Dixons had a reputation as a cheap stockist of electronics products with minimal customer service, the European equivalent of a cut-rate Best Buy. But Cook wasn't worried. "I talked to many people and John was the best by far," he responded in an email to one customer who wrote him. "I think you will be as pleased as I am. His role isn't to bring Dixons to Apple, [it is] to bring Apple to an even higher level of customer service and satisfaction."

But the concerned customers were right. Eschewing the Apple Stores' reputation for great customer service, Browett immediately set about trying to cut costs by reducing hires and staff hours. He fixated on meeting sales targets, which had never been Apple's top priority. The changes were almost unanimously met with complaints and disdain. Six months after Browett had started at Apple, Cook fired him. In a rare acknowledgment of error, spokeswoman Kristin Huguet said, "Making these changes was a mistake, and the changes are being reversed. Our employees are our most important asset and the ones who provide the world-class service our customers deserve."

Browett had been a bad fit for Apple. He was a sales-and-profit-oriented executive who didn't gel with the stores' laid-back, softly-softly philosophy.

Cook had made a rare error in judgment by choosing someone whose approach to business prioritized the bottom line. Browett also was contrite. "I just didn't fit with the way they ran the business," he later recalled. "It was one of those shocking things where you're rejected from an organization for fit rather than competency."

Browett's departure was an embarrassing black eye for Cook. It made him look clueless and out of touch, especially in his first year as CEO. Steve Jobs had made a couple of hiring mistakes over the years, but Jobs was better known for picking great collaborators—most notably Steve Wozniak and Jony Ive at Apple, and John Lasseter and Ed Catmull at Pixar. Cook didn't have these amazing collaborators on board, and his vision for Apple differed greatly from Jobs's. There was extra pressure on him, and the Browett hiring and firing made it look like he hadn't done his homework, or, worse, that his judgment was wrong. What was he thinking?

Even more significant than Browett's firing was the departure of Scott Forstall, once a potential candidate for the CEO position after Jobs's death. Forstall had started his career at NeXT and had risen rapidly through the ranks. He followed Jobs to Apple, where he was one of the major forces behind Apple's fabulously successful Mac OS X. He was rewarded for this success, later given the tricky task of developing the software for the original iPhone. His stellar performance earned a higher public profile, with Cook choosing him to show off Siri, Apple's new AI assistant, at the iPhone 4S keynote. He was moving up quickly, just like he had at NeXT. A 2011 *Bloomberg Businessweek* profile of him, written shortly after Jobs's death, called him the "Sorcerer's Apprentice," a "mini-Steve," and "the best remaining proxy for the voice of Steve Jobs." *Fortune* reporter Adam Lashinsky's book *Inside Apple* pegged him as a potential Apple CEO in waiting, ready to take over when the Tim Cook era was over. With his rapid rise at Apple, it seemed that he was one of the most likely candidates to succeed Steve Jobs. But by the end of 2012, he was gone.

Forstall was the least well liked of Steve Jobs's executive team, but he had a kinship with the Apple cofounder that protected him. According to Lashinsky, if there was a knock against Forstall, it was that he wore his ambition too openly on his sleeve. His hunger for power and influence within the company irked his fellow executives and coworkers, but he was favored by Jobs, who appreciated his drive and competence. When Jobs died, this safety net was removed. Forstall's departure in 2012 followed two subpar software launches: the disappointing launch of Siri and the disastrous introduction of Apple Maps. Siri began as a spin-out from the SRI International Artificial Intelligence Center and an offshoot of a DARPA-funded AI project. It launched as a third-party iOS app in 2010. Apple quickly acquired it and set about remaking it in its own image.

When Forstall introduced Siri onstage at the iPhone 4S keynote, it was the talk of the event. It delivered on a digital personal assistant concept that Apple had first shown off in 1987 under the name "Knowledge Navigator." But Siri was met with mixed reviews when users actually tried it. Steve Wozniak, Apple's cofounder, was a public critic of Apple's version, despite professing to be a big admirer of the original third-party app. He pointed out that Siri had originally won him over by understanding questions like "What are the five largest lakes in California?" and "What are the prime numbers greater than 87?" But when Apple launched its own version, Woz pointed out that his query about lakes now called up links to lakefront properties, and the question about prime numbers prompted responses about prime rib. Within a year of the launching, two of the executives behind the original Siri app—Adam Cheyer and former Siri CEO Dag Kittlaus—left Apple in order to pursue other projects. The pair cofounded another AI start-up called Viv, which was later sold to Samsung for $215 million, leaving the disastrous Siri launch behind them.

While Siri was viewed as a missed opportunity, it was nothing compared to the flop of Apple Maps. Apple's mapping software was announced during the Apple Worldwide Developers Conference (WWDC) on June 11, 2012. It shipped with iOS 6 and replaced Google Maps as the iPhone's default mapping service. Mapping was still in its infancy, but it was increasingly clear that maps were a killer app for smartphones. Maps and directions were a key feature, rife with potential revenue opportunities like mobile advertising. Apple knew it was too important to cede control to Google and so decided to take control and develop its own service. Apple Maps features included turn-by-turn navigation, 3-D maps, Flyover, and Siri integration, which would set it apart from the popular Google Maps.

Unfortunately, though, when Apple Maps launched on September 19, it didn't take users long to label it a broken mess. There were all sorts of problems: warping of landscapes so they looked like Salvador Dalí paintings, directing UK searches for London to "London, Ontario," rather than the United Kingdom's capital city, and dangerous issues like suggesting that users enter Fairbanks International Airport in Alaska by driving across one of the runways. The *New York Times* described Apple Maps as "the most embarrassing, least usable piece of software Apple has ever unleashed." And an Apple executive was reportedly half an hour late for a meeting because Apple Maps had directed him to the wrong place. Had Apple kept it simpler, there's a good chance that it would have worked as hoped. Apple had overreached. It had tried to launch a feature-rich replacement for Google Maps, which had been around for about seven years and had lots of mature features. But Apple insisted on taking on Google and was trying to do too much, and ultimately they paid the price.

As the executive in charge of Apple Maps, Forstall was ultimately responsible for the new application's failure. An article by *Business Insider*'s Jay Yarow, titled "The Apple Maps Disaster Is Really Bad News for Apple's 'CEO-in-Waiting,'" emphasized that this was Forstall's "second consecutive high-profile screw up with iOS software." The year before, he had pushed Apple to add Siri to iOS, with less than stellar results. Cook reportedly asked Forstall to make a public apology, but Forstall refused. Neither Cook nor Forstall have addressed the issue.

After the public outcry had been rumbling for two days, with no sign of a Forstall apology, Cook took the initiative himself. He sent out a public apology letter to Apple users, which detailed some of the problems with Maps and laid out a promise for the future:

> At Apple, we strive to make world-class products that deliver the best experience possible to our customers. With the launch of our new Maps last week, we fell short on this commitment. We are extremely sorry for the frustration this has caused our customers and we are doing everything we can to make Maps better.

> Everything we do at Apple is aimed at making our products the best in the world. We know that you expect that from us, and we will keep working non-stop until Maps lives up to the same incredibly high standard.

To some, Cook's apology was seen as a sign of weakness. According to *The Week* magazine, Cook's apology was "a form of abasement that has some people saying, 'That would never have happened if Steve Jobs were still alive.'" The editorial went on to ask, "Is Cook's apology proof, once and for all, that he's no Steve Jobs?" Steve Jobs had *never* apologized for Apple's mistakes. A couple of years prior, when users complained that the new iPhone 4 was dropping calls when their fingers made a connection across its exterior antenna—the so-called Antennagate scandal—Jobs preposterously said users were "holding it wrong." (He eventually held a press conference about Antennagate and offered customers a free bumper case to attenuate the problem, but he never made an explicit apology.)

But it appears Cook wanted to do things his own way. Seeing him as a disruptive influence on the team and a troublemaker who wouldn't take responsibility, he fired him. Cook's colleague Greg Joswiak obliquely praised the move, saying Cook had acted very decisively about making changes to the Maps organization, although he declined to discuss or name Forstall personally. Calling Cook "bold and decisive" and "every bit a leader," Joswiak said Cook "got very personally involved with what we were going to have to do to right that ship."

There appeared to be no protest whatsoever about Forstall's firing from his fellow executives, at least not publicly. Internally, his departure was reportedly "cheered" by staffers. One former Apple employee, who asked not to be named, said Forstall played a lot of politics, often took credit for others' work, pushed coworkers aside to get ahead, and caused a lot of friction. He was also nakedly ambitious, which alienated him from coworkers. "He wanted to be involved in too many things that weren't his business," the staffer said. "And I think that upset other people a little bit. . . . I think too many people were just a little fed up with him."

Tony Fadell, Apple's former senior vice president of the iPod division and one of the "fathers of the iPod," told the BBC that Forstall "got what he deserved." It was widely rumored that there was bad blood between Forstall and Fadell. The pair reportedly clashed constantly when both were working on developing the iPhone (Forstall was in charge of the software, and Fadell led the engineering). They clashed over resources, personnel, and credit until Fadell quit Apple in 2008. In the same interview, Fadell said that he thought Forstall's firing and Cook's reorganization of the executive team was a good move. "So, I think Apple is in a great space, it has great products and there are amazing people at the company, and those people actually have a chance to have a firm footing now and continue the legacy Steve left," he said.

Many employees thought that playing politics was the main reason for Forstall's firing. He was responsible for the software that powered the iPhone, and his star rose as the iPhone took off. He became very powerful internally, and Apple analyst Horace Dediu said he heard rumors that Forstall was starting to run his own projects—recruiting hardware engineers for his own projects—without involving Cook or Apple's other executives. "As much as he was contributing, as much as he was doing right, the capital offense at Apple is that you disobey and you overstep your boundaries."

Dediu said he's doubtful that Forstall got fired simply for screwing up the new Apple Maps venture, because Apple's culture tends to be forgiving of mistakes. If Forstall refused a direct order from Cook to make a public apology, it was this defiance that likely doomed him. Dediu said that "Tim felt that 'hey, I'm being tested. These guys are starting to exercise power. I need to be very decisive,' and I think partly his reasoning was like, 'I've got to do this public execution, in order for others to take me seriously.' And so from then on he may have had a lot less trouble" internally.

A few months after Forstall's ouster, Cook gave an interview to *Bloomberg Businessweek* that revealed a little more about the episode, and about Cook's leadership style. Cook said the management shake-up was to increase collaboration at the company—just as the press release had said. "We have an enormous level of collaboration in Apple, but it's a matter of taking it to another level," he said. "You look at what we are great at. There are many things. But the one thing we do, which I think no one else does, is integrate hardware, software, and services in such a way that most consumers begin to not differentiate anymore. They just care that the experience is fantastic. So how do we keep doing that and keep taking it to an even higher level? You have to be an A-plus at collaboration."

Cook didn't address the firing of Forstall and Browett directly, but what he said next about how the members of the executive team worked together and took responsibility was quite revealing. "The thing that ties us all is we're brought together by values. We want to do the right thing. We want to be honest and straightforward. We admit when we're wrong and have the courage to change. And there can't be politics. I despise politics. There is no room for it in a company. My life is going to be way too short to deal with that. No bureaucracy. We want this fast-moving, agile company where there are no politics, no agendas." Reading between the lines, Cook fired Forstall because he played politics and had his own personal agenda: He wouldn't admit he was wrong, he wouldn't apologize, and he showed an unwillingness to change. For Cook and his leadership style, these were cardinal sins.

Passed Peak Apple?

Unsurprisingly, not everyone thought firing Forstall was a good idea. "We've passed peak Apple: it's all downhill from here," Dan Crow wrote for the *Guardian* in late 2012. "The decision to dump Google's maps for its own, and the changes at the top of the company to eject Scott Forstall and John Browett point to a subtle downward trajectory." Although Apple continued to do extremely well financially (revenues for 2012 were $156.5 billion, of which a whopping $46.33 billion was profit), many pundits attributed this to Jobs and the fruits of his legacy. People couldn't resist the narrative that Apple was suffering a thousand tiny cuts that would bring it crashing to earth sooner or later.

Even Apple's own ad agency, TBWA\Media Arts Lab, seemed to buy into that narrative, much to Apple's chagrin. In January 2013, the *Wall Street Journal* published a story titled "Has Apple Lost Its Cool to Samsung?" The article suggested that Samsung, which was then aggressively marketing its new Samsung Galaxy S3 smartphone, had caught Apple in a slump. In private emails to Apple executives, later unearthed as part of the *Samsung v. Apple* patent trial, TBWA wrote, "We understand that this moment is pretty close to 1997 in terms of the need to help pull Apple through this moment." In 1997, when Jobs returned to Apple, he commissioned the famous, award-winning "Think Different" campaign to remind the world that despite Apple's financial woes, there was a great company waiting to emerge. It worked spectacularly, breathing new life into a company that was in serious trouble. In 2012, the situation seemed to have reversed. Apple was doing well financially, but the public perception was off—many worried that it was in decline.

But those at Apple had faith. Apple's marketing chief, Phil Schiller, fired back. "This is not 1997. . . . In 1997 Apple had no products to market. We had a company making so little money that we were six months from [going] out of business. . . . Not the world's most successful tech company making the world's best products[,] having created the smartphone and tablet form factors and leading in content distribution and software marketplaces. Not the company that everyone wants to copy and compete with."

Tellingly, TBWA never produced an updated "Think Different"–style ad for the Tim Cook era.

Cook Starts to Change Apple

While shuffling the executive ranks and increasing collaboration was a major internal adjustment, to observers on the outside little seemed to have changed. Cook seemed on the whole to be continuing the legacy of his predecessor. However, there were several hints of the way he would steer Apple in the coming years.

The first big change Cook made as CEO came in January 2012, just five months after taking over. He held an internal town hall meeting, following Apple's blockbuster earnings report for the previous quarter. He had told employees that the meeting would discuss "exciting new things"—and one of those things turned out to be Apple's philanthropic efforts. This was a significant departure from Steve Jobs's vision for Apple. Jobs had been famously stingy when it came to charities, arguing that the most charitable thing he could do was increase Apple's value so that shareholders had more money to give away to the causes of their choice. Under Jobs, Apple made no significant contributions to charity—at least not publicly. The only public philanthropy the company participated in was U2 frontman Bono's (PRODUCT)RED charitable effort, which raises money to help fight HIV/AIDS in Africa through selling branded RED products. Since 2006, the company has sold half a dozen different RED versions of the iPod and iPhone, raising more than $160 million.

But Cook holds a different view. He plans to donate his entire fortune to philanthropic projects, after paying for his nephew's college education. As CEO, he instituted a charitable matching program for employees, under which Apple would match donations up to $10,000 per employee per year. It was a huge hit. In its first two months, the company and its employees donated $2.6 million. At the town hall meeting, he revealed that Apple had also donated $50 million to Stanford's hospitals. Cook didn't spell out why Stanford was chosen, but the company and Steve Jobs have had a long association with the university and its medical facilities. Jobs delivered a famous commencement speech to the university in 2005 and had received treatment for cancer at the Stanford Cancer Center. Apple has also donated several hundred million dollars to various educational and environmental initiatives since 2011, including money donated to hurricane relief, wildfire recovery efforts, and flood relief in China, among many others.

As well as investing significantly in education schemes, Apple has made large donations to charities focusing on health and human rights. About a year and a half after Cook became CEO, he agreed to volunteer his time for an auction operated by Charitybuzz that would allow bidders the opportunity to enjoy a coffee with him at Apple's headquarters. The bid raised $610,000 for the Robert F. Kennedy Center for Justice and Human Rights, matching the Charitybuzz record set by a Lamborghini Aventador LP 700-4 Roadster just two months earlier. This was twice as much as any other auction had raised for the RFK Center, and more than twelve times as much as the auction's "estimated value" when it went live just a few weeks earlier. App developers, accessory makers, and entrepreneurs were among the wishful bidders, but the auction winner remained anonymous.

As 2014 drew to a close, the company made a whopping $20 million donation to PRODUCT(RED) to help fund its fight against AIDS. The money was raised through partnerships with app developers, who gave up some of their proceeds from sales and exclusive in-app purchases, and through Apple product sales during the two biggest shopping days of the year—Black Friday and Cyber Monday. "I'm thrilled to announce that our total donation for this quarter will be more than $20 million—our biggest ever—bringing the total amount Apple has raised for PRODUCT(RED) to over $100 million," Cook wrote in an email to Apple employees. "The money we've raised is saving lives and bringing hope to people in need. It's a cause we can all be proud to support."

The company hasn't released any new numbers regarding employee donations, so the amount raised since Cook's 2011 announcement remains unknown. But in 2018, after the Trump administration adjusted U.S. tax law to allow Apple to repatriate almost $250 billion of overseas cash, Cook announced that the amount of matching donations would be doubled at a rate of two to one instead of one to one. He sent Apple employees a note saying the company would double employee charitable donations (still topping out at $10,000 annually) through the end of the year. Apple would also double the amount matched for each hour their staff donated time.

Cook's new charitable stance was generally well received, but some noted that the money donated was a drop in the bucket and that it was somewhat at odds with the labor issues Apple faced in its own supply chain. "Apple's charity efforts fall short when you look at the $97.7 billion the company now has in cash," wrote reporter Sarah Mitroff in VentureBeat. The money "that Apple has given away is a small drop in the bucket even compared to $46.33 billion, the revenue Apple made in the first quarter of 2012. In addition, the labor issues stemming from Foxconn, the company that manufacturers the iPhone and other Apple products, only hurts the company's newly found charitable image." In his rocky first year, Cook was nonetheless making small changes that would foreshadow bigger positive changes at Apple later on.

Supply Chain Initiatives

This included making changes in Apple's supply chain practices. In February 2012, ABC aired a *Nightline* special on Foxconn, Apple's biggest manufacturer. Although it was filmed with Apple's approval, the special nonetheless shed a deeper and troublesome light on the conditions under which Apple's products were produced. Interesting details included the fact that it takes five days and a shocking 325 hands to build a single iPad. The special also revealed that Foxconn workers work twelve-hour shifts and pay $0.70 per meal, their six- to eight-person dorms cost $17.50 per month in rent, and they are paid $1.78 per hour for their work.

The *New York Times* also published a Pulitzer Prize–winning series of investigative reports about Foxconn's working conditions. Cook reacted with uncharacteristic outrage, sharing his thoughts with employees in a manner that couldn't have been more different from Steve Jobs, who had been criticized for being uncaring. In an internal email, Cook said he was "offended" and "outraged" by the report. He was direct and transparent with his employees, writing, "We care about every worker in our worldwide supply chain. Any accident is deeply troubling, and any issue with working conditions is cause for concern. Any suggestion that we don't care is patently false and offensive to us. As you know better than anyone, accusations like these are contrary to our values. It's not who we are." Reinforcing that Apple had been improving conditions for hundreds of thousands of workers, Cook was determined to set the record straight.

Apple almost immediately hired the Fair Labor Association, a Washington, D.C.–based group dedicated to ending sweatshops around the world. Apple charged the FLA to audit Foxconn factories in Shenzhen and Chengdu, China. The move was seen as a significant step toward Apple's cleaning up and being responsible for its supply chain, one of the six values outlined on its website. Apple was the first tech company to employ the FLA, which had built a reputation reforming the supply chains of the food and clothing industries. Until October 2016, when Apple stopped working with the FLA, it was the only tech company to do so.

At a Goldman Sachs conference shortly afterward, Cook bluntly addressed the charges of worker abuse in the supply chain. In his keynote speech, he said that Apple would not rest until every worker was guaranteed a safe working environment without discrimination and at a competitive salary. He said any suppliers that didn't take care of their workers would have their contracts terminated. "Apple takes working conditions very seriously, and we have for a very long time," Cook said. "Whether they are in Europe, Asia or the United States, we care about every worker." He spoke from his own blue-collar personal experience. "I've spent a lot of time in factories personally, not just as an executive," he said. "I worked in a papermill in Alabama and an aluminum factory in Pennsylvania; I care, and we have hundreds of employees who work in our factories full time. They care, too. We are very connected to worker conditions on a granular level."

Cook's pledge of meaningful reform was generally well received. Dara O'Rourke, associate professor of environmental and labor policy at the University of California, Berkeley, told the *New York Times* that Cook deserved kudos for speaking up. "I want to give credit to Tim Cook for this," he said. "He's admitting they've got problems." Daniel Diermeier, a Northwestern University professor specializing in reputation management, agreed. He said the bad press had clearly prompted Cook to take action, but that he was in a good position to make positive change. "I think he probably has a deeper understanding, and this is more personal for him than it might be for other executives."

But there were some who doubted Cook's claims. Jeff Ballinger, a labor activist and researcher, said Cook's promises of reform were questioned. "It looks like a pattern I've observed before," he said. Cook is hoping "things will die down. It's not very convincing."

But Cook was determined to make changes. He invested time and effort in working conditions, making a point of visiting Apple's factories and interacting with the workers there. At the end of March, he flew to China to take a tour of a new Foxconn assembly plant in Zhengzhou that employed around 120,000 workers, many of whom assembled the iPhone. Apple released a photo of the visit that was reproduced around the world. It was a big deal that Cook was getting more involved in supply chain logistics—Steve Jobs never had his picture taken on an assembly line—but a lot of cynical bloggers dismissed it as a photo op.

Despite Cook's pledges to improve worker conditions, more than two thousand workers rioted at a Foxconn plant in Chengdu over a minor incident, reportedly a theft in one of the factory's dormitories. Tensions at the plant were high over conditions and pay. Not long afterward, a twenty-three-year-old Foxconn worker in Chengdu committed suicide by jumping from his apartment. He'd only been with Foxconn for a month, and it came hot on the heels of Apple's promises of reform.

It wasn't the first suicide at Foxconn. There had been one death in 2007 and another in 2009, but in 2010 there was a sudden massive upsurge, with an estimated eighteen employees attempting suicide and at least fourteen deaths. The first of these took place in January 2010, when a young factory worker named Ma Xiangqian jumped to his death. Ma had recently been demoted to cleaning toilets after accidentally breaking some factory equipment. He had been working triple the legal overtime limit. "Life is hard for us workers," said his sister, Ma Liqun, shortly following Ma's death. "It's like they're training us to be machines." Sun Dan-yong, a twenty-five-year-old Foxconn employee who died in July 2009 after throwing himself from an apartment building, did so after losing an iPhone prototype in his possession. Prior to his death, he claimed he was beaten and his residence searched by Foxconn employees.

Foxconn chairman Terry Gou initially responded with shocking indifference, believing that "a harsh environment is a good thing." But when the fourteenth Foxconn employee leapt to his death in May 2010, Foxconn started taking suicide prevention measures. As a first step the company erected more than three million square meters of yellow mesh netting around its buildings to catch jumpers, hardly tackling the root of the problem, but it also increased wages for the factory workers in Shenzhen by 30 percent to 1,200 renminbi ($176) per month, and promised a second raise six months later. Finally, it set up a twenty-four-hour counseling center staffed by one hundred trained workers, and opened a special stress room where workers could use baseball bats to take out their frustrations on mannequins.

The suicides were quickly linked to Apple. Although Apple was not the only large company that used Foxconn, it was the biggest and best known. The scandal also seemed to contrast most strongly with Apple's progressive image. The authors of the book *Becoming Steve Jobs*, which paints Apple and Jobs in a fairly positive light, ask, "How could a company with Apple's cherubic marketing glow make its devices in Foxconn factories where the drudgery and difficult working conditions resulted in more than a dozen assembly-line workers committing suicide?"

But Steve Jobs was the wrong person to speak up on the subject. Like Terry Gou, Jobs himself was not averse to promoting a tough work environment. When he defended Foxconn after the 2010 suicides, he said its factories were actually "pretty nice" and "not a sweatshop." The line that came over worst, however, was his comment, "We're all over this," which struck many people as uncaring and insensitive.

Nonetheless, Apple did make changes, which accelerated significantly after Cook took over as CEO. The Fair Labor Association released its first report, issued in August 2012, identifying 360 "remedial action items" that needed fixing regarding worker safety, pay, and conditions. The items included things like pay and hours worked, overtime, health and safety training, unemployment insurance, child labor, and ending an exploitative internship program.

There were some problems that had remained unresolved in the seven months since Cook hired the FLA to work with Apple. The FLA report noted that some of the biggest action items had not been fixed, including union representation and reducing workweek hours to follow Chinese labor law. But reducing hours and overtime wasn't popular with many workers, who often preferred to work longer hours and accrue more pay, which was saved or sent home.

But overall, the FLA noted that Apple and Foxconn had made significant progress in reforming worker conditions. With deadlines stretching up to fifteen months in some cases, the FLA report indicated that Apple and Foxconn had already implemented 284 of the recommended changes ahead of time. "Our verification shows that the necessary changes, including immediate health and safety measures, have been made," said FLA CEO Auret van Heerden in a statement. "We are satisfied that Apple has done its due diligence thus far to hold Foxconn accountable for complying with the action plan, including the commitment to reform its internship program." In his first year as CEO, Tim Cook had already made more improvements to supplier responsibility than Steve Jobs had in his entire time at Apple. He wrote in an email to employees in early 2012, "No one in our industry is driving improvements for workers the way Apple is today."

In the years since, Apple has worked to improve its supply chain, although it has still received occasional criticism from labor rights activists and other organizations. Given its power and profit margins, many argue that it could be and should be doing a lot more, and the conditions in supplier factories are still pretty miserable. Li Qiang, director of China Labor Watch, criticized Apple for keeping suppliers' margins low, which in turn depresses wages. Apple's suppliers only make 5 to 10 percent margins, he said, which isn't enough to raise workers' wages. "If they really wanted to change labor, they should be paying more," he said. "Ultimately, it's not the supply chains or the factories themselves, but it's Apple themselves who are not willing to pay more."

He also questioned Apple's claim that 90 percent of its factories adhere to its rules about excessive overtime, which under Chinese labor law limits the working week to sixty hours. He said workers are effectively forced to work overtime to make up for their low wages. Representatives from his group went to a Pegatron factory in Shanghai and examined about a thousand pay stubs. They found that 70 to 80 percent had worked more than sixty hours, he claimed. When they reported their findings to Apple, he said the company dismissed their concerns because the sample size was too small.

Many activists believe that the problem is with the manufacturing system itself. Ted Smith, founder and former executive director of the Silicon Valley Toxics Coalition and chair of the Electronics TakeBack Coalition, thinks the solution is for Apple to build its own factories and directly employ the line managers and workers. "The scale of what they're trying to do or what they are doing is . . . almost beyond comprehension," he said. "Even if they have the best people in the world, in positions of responsibility and influence and power, within headquarters, trying to manage 756 factories with literally millions of workers, or facilities . . . is just incredibly daunting."

Smith said his group started lobbying Apple just before Jobs died. He used to feel, under Jobs, that Apple didn't care, or was too stingy to make any changes. He praised Cook for bringing in top people, "again, a huge change from what it used to be," especially Cook's hiring of Lisa Jackson, the former head of the EPA. He also praised the academic advisory board established to study workforce problems (see page 110) and Apple's willingness to be public and make more disclosures through its annual supplier responsibility reports. "I have seen tremendous improvement . . . and I attribute that mostly to Tim Cook's leadership," he said.

Jeff Ballinger, a veteran antisweatshop activist, writer, and political scholar, agrees that the problem is the system itself. Apple wouldn't have so many problems with the supply chain if it didn't contract out its manufacturing. "I would like to see vertical reintegration," he said. "Why can't these companies make things themselves? The abuse comes about because the paradigm is you put six contractors, you put a bid out and take the lowest price." Because of this, it's impossible for Apple to go back to the contractors and say, 'Can't you give them time off, pay bonuses, etc.?'" He said monitoring was ineffective. "The real issue is the system," he said. "Monitoring is a dodge."

Jenny Chan, a professor at Hong Kong Polytech and former SACOM (Students And Scholars Against Corporate Misbehaviour) organizer, and author of *Dying for an iPhone*, said Apple is aware of many of the safety issues and doesn't do enough, and sometimes doesn't do anything at all. She advocates for more labor organization, from the bottom up, to empower the workers. "Workers should participate in trade union elections," she said. "Workers should be represented in the occupational safety and health committees. So workers should have their voices and their opinions, their decisions or their discussions being taken more seriously. But at the moment I do not see the structure in place."

Heather White, a New York–based documentary filmmaker of *Complicit*, a film about terrible working conditions in Apple's supply chain, took the hardest line of all. She questioned why Apple is in China at all. It's a repressive, corrupt regime with a terrible record of fair labor practices. "If any of these corporations in the electronics industry are serious about complying with their own codes of conduct which they post on their websites and claim are representative of their commitments to all of their stakeholders, they would have to start talking about leaving," she said. She admitted that wasn't realistic, but if Apple and Tim Cook were serious about its code of conduct, the company would encourage freedom of association and worker rights and really crack down on health and safety standards.

But Deirdre O'Brien, Apple's Veep of Peeps, who for many years worked closely with Cook in operations, defended the company's progress on advancing worker rights. "I actually feel like that's one of the best things that Tim actually has done," she said. "I really think the work that Apple's done in the supply chain is exceptional and we've worked to be a true leader in that space. So instead of ignoring problems and saying, 'That's somebody else's issue,' he stepped right up and said, 'We want to be a leader here and let's talk about these things and let's go to work.' . . . We are very focused on ensuring their well-being and there are the education opportunities, safety issues, all those things we take incredibly seriously."

And Cook has created new initiatives to help move Apple in the right direction. Right after WWDC, in July 2013, Apple formed an academic advisory board to oversee its supplier responsibility program. The board was a natural extension of Cook's commitment to reforming the supply chain. Initially made up of eight professors representing top American universities, and chaired by professor Richard Locke of Brown University, the board was part of Cook's efforts to improve labor practices in the supply chain. Their purpose was to conduct or commission research on labor standards in Apple's supply chain, share existing research that may improve practices and policies, and make recommendations to Apple about positive steps that it may take. "I hope that the Board will . . . shape the practices of Apple and its suppliers so that all employees involved in Apple's supply chain . . . are paid living wages, work within the legal work hour regimes, work in environments that are safe and where they can express their rights as citizens," Locke said in a statement.

Each year, Apple has also sought to expand its supplier responsibility report, first published in 2005 along with the Apple Supplier Code of Conduct. In 2007 Apple audited only 39 facilities as part of the program. This rose to 102 in 2009 and to 756 across thirty countries in 2017. It also established guidelines for areas such as dormitories, juvenile worker protections, medical nondiscrimination, pregnancy nondiscrimination, prevention of involuntary labor, wages and benefits, and working hours.

Under Cook, Apple has taken a more hands-on role in launching initiatives targeting workers. In 2017 the company launched a new health awareness program intended for women working at its suppliers in India and China, offering access to services and education on self-examination for early cancer detection, nutrition, personal care, and maternal health. Jeff Williams said that by 2020, Apple hopes that this program will have reached one million women.

Apple's financial muscle also means that it is able to dictate many of the terms of business to its suppliers. In 2018, Apple forced one of its suppliers in the Philippines to repay a total of $1 million it had charged for recruitment fees for factory jobs. Such steps, which could only be made by a company with the heft and leverage of Apple, are viewed as crucial. Though many companies use overseas manufacturers and suppliers for their components, Apple is the most immediately recognizable one. And as the Foxconn suicides story showed, many people draw no distinction between Apple and the companies it hires to make its products.

This approach to ensuring suitable behavior among those who work with Apple goes even further than direct suppliers. According to a 2012 report in Engadget, Apple also insists on similar ethical behavior from companies that make official MFi (Made for iPhone) accessories for Apple's handsets. As the Engadget journalist wrote at the time, "Pushing [this edict] out to the larger accessory ecosystem would be a concrete example of Apple using its 800-pound-gorilla status in the consumer electronics space to influence more companies to behave ethically on worker rights, environmental issues and more."

Under Cook's leadership, Apple has managed to undo some of the damaging reports about its supply chain. In May 2014, the international aid organization Baptist World Aid Australia ranked Apple as the second-best company for improving working conditions for employees in its *Electronics Industry Trends Report*. Apple scored a B-plus, placing it just below Nokia. "Apple's inclusion in the top tier may come as a surprise given the public attention it has received for poor working conditions and child labour at Chinese suppliers like Foxconn and Pegatron," the report stated. Nonetheless, the report highlighted how Apple was changing for the better.

Success on the Horizon

Cook's first year as CEO wasn't without its challenges. Criticisms over Apple's relationship with Foxconn, whose labor conditions were terrible, came to fruition. Cook was questioned for some bold hiring and firing decisions and criticized for the launch of disappointing, underperforming products. But his response to the Foxconn fiasco was generally received as admirable, as he placed increasing resources on cleaning up the supply chain, and his efforts to tighten Apple's policies on privacy after the Path breach (the Path app secretly uploaded users' data) were well regarded. He also overcame initial concerns about new Apple products being second-rate with the launch and smashing success of the new iPhone. In September 2012, Apple introduced the iPhone 5, the company's first iPhone to be released after Steve Jobs's passing. The iPhone 5 boasted major design changes. It had a glass body and a taller screen, and Lightning, Apple's new compact dock that replaced the 30-pin connector found on every iPhone before it. This change caused some controversy at the time, but the success of the iPhone 5 despite this was undeniable.

When Apple began taking preorders on September 14, more than two million units were sold within the first twenty-four hours. Apple claimed that the number of preorders was more than double the previous record of one million held by the iPhone 4S. When the device went on sale, opening weekend sales topped five million units, exceeding the four million the iPhone 4S had managed the previous year's opening weekend. If anyone was worried about Apple's future, it wasn't reflected in iPhone sales, which only climbed to greater heights under Cook's watch.

That October marked one year since Steve Jobs had died. In a message posted to the homepage of Apple's website, Cook wrote, "I'm incredibly proud of the work we are doing, delivering products that our customers love and dreaming up new ones that will delight them down the road. It's a wonderful tribute to Steve's memory and everything he stood for." Cook was keeping Jobs's legacy alive. Apple had not failed after Jobs's death, as many had expected—Cook was actually bringing the company to new heights. And the world took note. In December 2012, Cook was named one of *Time* magazine's "100 Most Influential People" in the world. In a story for the article, former vice president Al Gore, who had served on Apple's board of directors since 2003, wrote:

It is difficult to imagine a harder challenge than following the legendary Steve Jobs as CEO of Apple. Yet Tim Cook, a soft-spoken, genuinely humble and quietly intense son of an Alabama shipyard worker and a homemaker, hasn't missed a single beat. Fiercely protective of Jobs' legacy and deeply immersed in Apple's culture, Cook,

51, has already led the world's most valuable and innovative company to new heights while implementing major policy changes smoothly and brilliantly. He has indelibly imprinted his leadership on all areas of Apple—from managing its complex inner workings to identifying and shepherding new "insanely great" technology and design breakthroughs into the product pipeline.

Gore hit the nail on the head. He knew Cook better than most, and he knew him to be a fantastic leader. Cook's place on the *Time* list showed that despite a rocky start to his CEO ship he was the right person to lead Apple into the future while still keeping its legacy intact.

All year, Apple's stock had been climbing, largely because of record sales of the iPhone. Less than a month after Jobs's death, Apple stock hit a new all-time high of $413, boosting the company's market cap to $390 billion. It was the beginning of a dizzying upward climb of Apple's stock under Cook. Analysts pointed out that with any luck it would soon be worth more than Google and Microsoft combined. By the end of January 2012, AAPL was valued at $447.61 and passed ExxonMobil to become the most valuable publicly traded company in the world.

In February 2012, after strong quarterly earnings, Apple's share price hit a new high point of $500. AAPL had risen $75 per share—nearly 18 percent—in less than a month. Only a month after that, it hit $600. In August, almost exactly a year after he took over as CEO, Apple shares soared to a record price of $665.15 and a market capitalization of $622.98 billion. This was the highest market capitalization ever reached by a publicly traded company and surpassed a record set by Microsoft in 1999. It was clear that the iPhone was a runaway smash that wasn't likely to end anytime soon, and Wall Street took note. Apple was now the biggest public company in history.

Chapter 7

Finding His Feet with Hot New Products

I n 2012, Cook and Apple were riding high. But they endured a bittersweet start to 2013. Despite its reporting record profit of $13.06 billion—the second-highest ever earned by a U.S. corporation—on the back of strong iPhone and iPad sales, shares in Apple plunged 12 percent. Investors were growing wary of competition from Android and worried about the company's prospects for growth. After all, it's hard to grow a company the size of Apple. According to the "law of large numbers," Apple would have to increase quarterly sales by tens of billions of dollars to grow just a few percentage points. By contrast, much smaller companies only have to increase revenues by a few million dollars to grow the same percentage. In just four months, Apple's valuation had slipped to $424 billion, almost a year after it overtook ExxonMobil to become the world's most valuable publicly traded company. "I don't like it either," Cook told disappointed shareholders during Apple's annual meeting in February after its share price dived 30 percent in five months. "Neither does the board or management . . . but we're focused on the long term."

Cook was determined to keep Apple at the cutting edge of innovation and pursue opportunities in new markets and new partnerships. Apple was still the world's most valuable company, and Cook planned to keep it that way by focusing on China, one of the fastest-growing smartphone markets in the world. Under his leadership, Apple began investing heavily in China, launching a new online store, striking deals with Chinese carriers, and opening new retail stores.

During a trip to China in early January, which saw visits with government officials, business partners, and Apple employees, Cook told the state-run Xinhua News Agency that Apple planned to open more Apple Stores in China. At the time, Apple had just two stores in the country—the first opened in Beijing in 2008, and a second was added in Shanghai in July 2010. Referring to China as a "very important country to us," he said, "We have some great sites selected, our manufacturing base is here, and we have incredible partners here."

During the same trip to China, Cook met with Xi Guohua, chairman of China Mobile, the world's largest carrier. Steve Jobs had had numerous secret meetings with China Mobile before his death, but Apple still hadn't reached a deal for them to distribute the iPhone. The iPhone was already being sold by China Mobile's rival, China Unicom, but Cook stepped up efforts to make the device more accessible to China Mobile's more than seven hundred million subscribers. Apple finally confirmed a deal in December 2013, and a month later, China Mobile began selling the iPhone 5S and iPhone 5C. This was a huge deal for Apple. After accounting for just 2 percent of the company's revenue in fiscal 2010, Jobs's final full year as CEO, China revenues under Cook exploded in just two years. Between 2010 and 2012, China accounted for more than $20 billion in new revenues, a jump of more than 600 percent. In just two years, China made up more than 12 percent of Apple's total revenues. But Cook believed it could be even bigger—and that the People's Republic would become Apple's biggest market—with significant investment in new retail stores.

However, it wasn't long after landing this big Chinese deal that Apple was forced to issue an apology to customers there after months of negative press surrounding its warranty practices. Chinese fans were unhappy that Apple was fixing broken iPhones still covered by a warranty, rather than replacing them with new or refurbished devices, which was already common practice in other markets. The company was also under fire for offering a ninety-day warranty on replacement components, despite a local law that stated the warranty should last for a full year. The *People's Daily* newspaper called Apple "empty and self-praising" in an article criticizing this policy, as well as Apple's refusal to speak to the Chinese press. A letter signed by Cook in early April assured fans that Apple was aware that a "lack of communications . . . led to the perception Apple's attitude was arrogant and that we do not care and attach importance to customer feedback." Cook continued, "We express our sincere apologies for any concerns or misunderstandings this gave consumers." Apple changed its warranty policy in China so that customers with a faulty iPhone would receive a new handset, rather than a repair, and Cook pledged to educate local resellers of the change to avoid any repair disputes. It was another example of Cook apologizing and owning up to problems. He demonstrated admirable leadership qualities to his employees, owning up to mistakes and problems that others, including his predecessor, did not.

Tax Dodging

The Chinese warranty fiasco wasn't the only public outcry Apple suffered in early 2013.

May brought new interest in Apple's tax practices, with a U.S. Senate subcommittee raising questions about the company's mammoth cash hoard overseas. Apple was accused of setting up offshore subsidiaries for large portions of its profits in an effort to "dodge" billions of dollars in U.S. taxes on $44 billion in overseas earnings over four years. Apple had paid almost $6 billion in U.S. taxes in 2011, but was accused of shifting $36 billion in taxable earnings away from the United States in 2012, avoiding a payment of $9 billion. "Apple wants to focus on the billions in taxes it has paid," said Carl Levin, then chairman of the subcommittee. "But the real issue is the billions in taxes it has not paid."

At this time, Apple held around $102 billion of its more than $145 billion in cash in foreign countries. Stashing most of its savings overseas looked a lot like tax dodging, and was very problematic to lawmakers. Testifying at a hearing by the Senate Permanent Subcommittee on Investigations, Cook insisted that Apple pays "all the taxes we owe—every single dollar." He insisted that Apple did not "depend on tax gimmicks. We don't move intellectual property offshore and use it to sell our products back to the United States to avoid taxes. We don't stash money on some Caribbean island. We don't move our money from our foreign subsidiaries to fund our U.S. business in order to skirt the repatriation tax."

Instead, the company had "real operations in real places with Apple employees selling real products to real customers." Cook testified that Apple's foreign subsidiaries held "70 percent of our cash because of the very rapid growth of our international business. We use these earnings to fund our foreign operations . . . [and] acquire equipment to make Apple products and to finance construction of Apple retail stores around the world." He agreed that the tax code was outdated, especially compared to those around the world. It was "very expensive to bring . . . cash back to the United States," and Apple and other American corporations were suffering "in relation to our foreign competitors, who don't have such constraints on the free movement of capital."

The committee had acknowledged that Apple had not broken any laws, but it claimed the company used the current tax code—which was criticized for not having "kept pace with the advent of the digital age and the rapidly changing global economy"—to its advantage, just as many other U.S. corporations, like Microsoft, Google, and Oracle, had been doing. Under the old code, companies like Apple paid up to 35 percent in taxes if overseas cash was repatriated. In December 2017, the Trump administration reduced that to a one-time rate of 15.5 percent on cash and 8 percent on other assets.

Apple's then chief financial officer, Peter Oppenheimer, who accompanied Cook at the hearing, used his opening remarks to highlight Apple's contributions to the U.S. economy—six hundred thousand domestic jobs and a plan to spend $100 million in late 2013 to open a Mac assembly facility in Texas that would use components manufactured in Illinois and Florida and equipment produced in Kentucky and Michigan. Apple was doing important work at home, and Oppenheimer wanted to make sure the committee knew that.

Though Apple was under extreme scrutiny from the committee, some members felt Apple was being treated unfairly. One, Senator Rand Paul, accused his colleagues of trying to "vilify" Apple and said the committee should apologize for forcing the iPhone maker to sit through a "show trial." He sympathized with Apple for having to work with "a bizarre and Byzantine tax code." Paul made it clear that he was "offended by a $4 trillion government bullying, berating and badgering one of America's greatest success stories."

The subject of taxes reared its head again in late May, when Cook made his second appearance at the AllThingsD conference. He reminded attendees that Apple had paid more than $6 billion in taxes—more than any other company in the United States—but admitted that it might "wind up paying more" if changes were made to close certain loopholes. When asked about testifying before Congress, he said that he "thought it was very important to go tell our story and to view that as an opportunity instead of a pain in the ass."

At AllThingsD, Apple's share price was also touched upon, and Cook admitted that its fall had been "frustrating for investors and all of us." So frustrating, in fact, that Cook docked himself $4 million in pay in 2013, while every other executive's compensation increased, after he decided that the CEO's salary should be aligned with company performance. In a proxy statement filed with the SEC, he said his decision was the result of a "strong desire to set a leadership example in the area of CEO compensation and governance," and a commitment to "including performance criteria in a portion of the equity awards [Apple] grants to executive officers in the future." It was a rare example of ethical leadership from a corporate CEO, who could have easily exploited his position of power instead. And Cook did fine regardless. Even with the pay cut, he still made plenty of cash, securing $2.8 million in bonuses on top of his $1.4 million in base salary.

Despite the negativity surrounding share prices, Cook maintained a positive outlook. He was keen to point out at AllThingsD that the fall in Apple share price was "not unprecedented. The beauty of being around for a while is you see many cycles." To assure the public that Apple was on the upswing, he reiterated plans for "incredible" new devices and said Apple had "several more game changers" up its sleeve. He also said that while he was very different from Steve Jobs, and had made some important changes, the core culture of Apple was the same. He reassured fans and investors that the culture born under Jobs, along with many of the same key people who brought us the iPhone, iPad, iPod, and the Mac, was still alive and well. Cook's performance was well received—the *Guardian* newspaper said his "preternatural calm" was "admirable"—but it was billionaire investor Carl Icahn who caused Apple stock to quickly climb 5.6 percent in August 2013. After what he described on Twitter as a "nice conversation" with Cook, during which the two discussed a larger stock buyback program, Icahn tweeted of his investment company, "We currently have a large position in Apple. We believe the company to be extremely undervalued." This tweet alone added nearly $12.5 billion to Apple's market value in just one hundred minutes.

Mac Pro + iOS 7

While under fire for falling share price and tax evasion, Cook's teams had been busy working on new products and software. Cook took to the stage at Moscone West in San Francisco in June 2013 for his second Worldwide Developers Conference, during which Apple unveiled OS X 10.9 Mavericks and improvements to the MacBook Air.

Though Cook appeared a little more comfortable onstage than he had at his first Apple conference as CEO, the products unveiled were received in a mixed light. The all-new Mac Pro, which packed high-end components into a compact aluminum cylinder, earned the nickname "trash can Mac" among some fans. The redesign divided the Apple community; while many marveled at its engineering, others criticized the lack of expandability. Its unusual cylindrical body made it difficult to upgrade with extra internal hard drives or amped-up video cards.

WWDC 2013 also offered Apple fans their first preview of iOS 7, an update significant for its complete redesign, which was led by Jony Ive following the departure of Scott Forstall and Cook's subsequent executive shuffle. The skeuomorphic look Jobs was so fond of, which made apps and icons look like real-world objects, was pushed aside in favor of cleaner, "flatter," and more modern visuals. "We understood people had already become comfortable with touching glass," Ive said in an interview with *USA Today* after iOS 7 made its public debut, "so there was an incredible liberty in not having to reference the physical world so literally." He added, "We were trying to create an environment that was less specific. It got design out of the way."

Cook called iOS 7 "a stunning new user interface," but his opinion wasn't shared by all iPhone and iPad users. Some labeled iOS 7 ugly and confusing, while others criticized its "shockingly basic" and "childish" visuals. Cook and Ive refused to waver, and the foundations laid by iOS 7's redesign—along with many of its original app icons—remain alive in Apple's mobile operating system today.

iPhone 5S Sets Records

Cook scored another public relations win in the fall with the release of new hardware and software. iOS 7 made its public debut in September, just one week before Apple launched the iPhone 5S, which combined the much-loved physical design of the iPhone 5 with all-new technologies. This included Touch ID, the new fingerprint recognition system that would change the way we secure iOS devices for years to come, and the A7 processor, Apple's first mobile chipset with a "desktop class" 64-bit architecture. The A7 took competitors by surprise, with Qualcomm, the largest manufacturer of mobile chips for rival Android-powered devices, almost two years away from launching its first 64-bit Snapdragon processor. The A7's redesigned architecture delivered unprecedented performance for a smartphone, with benchmark scores that easily surpassed those of competing devices from the likes of Samsung and Motorola.

These improvements helped make iPhone 5S the most exciting smartphone launched by Cook's Apple up to that point, and this was reflected in its sales performance. With a little help from the iPhone 5C, a more affordable alternative to the 5S, which offered slightly older specifications in a colorful plastic design, Apple sold a then-record-breaking nine million handsets over its launch weekend. The iPhone 5S sold three times as many units as the 5C, with demand far exceeding Apple's initial supply of devices and some fans having to wait more than a month for their orders to arrive. Six months after the iPhone 5S made its debut, Apple revealed that sales of the iPhone brand had exceeded half a billion units. The device even outsold Samsung's newly released Galaxy S5 by 40 percent in May 2014, despite being eight months old by that point.

A Good End to the Year

Though the beginning of 2013 had started off rocky with a significant fall in share price, the holiday sales period at the end of the year saw Apple achieve record revenue of $57.6 billion, with a net profit of $13.1 billion. Fifty-one million iPhones, twenty-six million iPads, and 4.8 million Macs were sold over the holiday period, and Apple's other products continued to enjoy impressive growth. "We are really happy with our record iPhone and iPad sales, the strong performance of our Mac products, and the continued growth of iTunes, Software, and Services," Cook said in a statement.

Cook bid farewell to 2013 with a memo to Apple employees that celebrated their efforts and reminded them of all the things Apple had achieved over the past twelve months. "As many of us prepare to celebrate the holidays with our loved ones, I'd like to take a moment to reflect on what we've achieved together over the past year," it read. "We introduced industry-leading products in each of our major categories in 2013, showing the breadth and depth of innovation at Apple." He continued, "Together we've shown the world that innovation at Apple goes beyond our products to the way we do business and how we give back to our community."

Cook highlighted the tens of millions of dollars Apple helped raise for important charities and relief efforts over 2013, and its ongoing support for (PRODUCT)RED, which was boosted by a series of one-off products designed by Jony Ive, all finished in red. Cook also teased "a lot to look forward to in 2014, including some big plans that we think customers are going to love." He concluded, "I am extremely proud to stand alongside you as we put innovation to work serving humankind's deepest values and highest aspirations. I consider myself the luckiest person in the world for the opportunity to work at this amazing company with all of you."

WWDC—iOS 8 and a Health Push

WWDC 2014 brought the introduction of iOS 8, which made significant improvements to Apple's mobile operating system. A health app, combined with the HealthKit framework, kicked off what would become a major push into the trillion-dollar healthcare industry, which Cook would later describe as a "huge problem in the world" that desperately needed attention. "With health care, there is a wide-open field to make some really profound contributions," he highlighted in a September 2014 interview with Charlie Rose. HealthKit is "yet another way to begin to build a comprehensive view of your life, which should empower you to take care of yourself over time," he continued. "And when you need help, it empowers you to take certain data to your doctor and get help from them." Health is obviously very important to Cook, a lifelong fitness fanatic, and under his watch Apple was soon to make its biggest push into health and wellness yet.

Other highlights from WWDC included OS X Yosemite, new color options and a cheaper price tag for the iPod touch, and a more affordable iMac priced at $899. Apple also unveiled Swift, its own programming language for Mac and iOS apps that's designed to be easier to learn and use than the aging and complex Objective-C. Apple would later make Swift open source, allowing anyone—even its rivals—to develop with and support the language, and contribute to its evolution.

Cook, who learned coding in college, has spoken passionately about the importance of getting kids to learn to program. "I think if you had to make a choice, it's more important to learn coding than a foreign language," he said. "I know people who disagree with me on that. But coding is a global language; it's the way you can converse with 7 billion people." On another occasion, he said, "Coding gives people the ability to change the world and from my perspective, it's the most important second language and the only language that is global."

Cook has been passionate about promoting people to learn to code (he'd later help launch an educational program called Everyone Can Code— see chapter 10). "We try to continually push ourselves to do more and more, not just on the hardware side but also in terms of developers' tools so they can take advantage of the hardware that's there, in the best way," he said in an early 2017 interview with the *Independent* about the 2014 introduction of Swift. "That's the heart of what the coding software Swift is all about. We've created the language and our hope was that you can get a lot more people coding, and then secondly have people push more to take advantage of the latest hardware." This emphasis on coding also helps Apple. Swift is becoming a popular way for developers to create apps for

Apple's platforms. And the more apps on iOS and Mac, the better it is for Apple.

Angela Ahrendts

In October 2013, after a year without a senior vice president of retail, Cook hired Angela Ahrendts to replace John Browett, who had been ousted alongside Forstall. Ahrendts, previously the CEO of Burberry, became the first woman to serve on Apple's executive team. Despite telling Cook during her interview that she was "not a techie," Ahrendts secured the position because she shared Apple's "values and our focus on innovation," Cook wrote in an email to Apple employees. "She places the same strong emphasis as we do on the customer experience. She cares deeply about people and embraces our view that our most important resource and our soul is our people. She believes in enriching the lives of others and she is wicked smart." Cook continued, "Angela has shown herself to be an extraordinary leader throughout her career and has a proven track record." Ron Johnson, former Apple retail chief, who pioneered the Apple retail store and the concept of the Genius Bar alongside Jobs, told Bloomberg that Ahrendts was a "terrific choice."

After transitioning from Burberry in the spring of 2014, Ahrendts set about integrating Apple's physical and digital retail businesses for a more seamless customer experience. She also launched a mission to breathe new life into Apple Stores around the world by turning them into communities focused on enriching the lives of their visitors. The stores were already wildly successful, but under Ahrendts there are far more events and classes than there used to be—and the stores are becoming places for people to meet and hang out. Under her leadership, employees are hired not for their sales experience, but for their empathy and compassion. Creative Pros have been introduced to help the Apple Stores' more than five hundred million daily visitors make the most of their devices and learn new skills. "We think of Apple retail [stores] as Apple's largest products," Ahrendts said during her first Apple keynote appearance in September 2017. "It's funny—we actually don't call them stores anymore. We call them 'town squares,' because they're gathering places where everyone is welcome." Cook added at the same event, "Apple retail has always been about more than selling. It's about learning, inspiring and connecting with people." He was determined to make Apple more accessible and easy to use for all, and his hiring of Ahrendts was a move that paid off.

Tim Cook's Tim Cook

Cook made other good personnel decisions that have contributed to Apple's continued success. In December 2015, he promoted his longtime lieutenant in operations—Jeff Williams—to the role of Apple's chief operating officer. Williams has been called "Tim Cook's Tim Cook." He's in charge of the operations side of the business, just as Cook used to head up operations when Steve Jobs was CEO. There are also an uncanny number of similarities between Williams and Cook.

Where Jobs and Cook worked well as a partnership in part because they were so different, Williams and Cook work well because they're so alike. "Williams is in many ways a doppelgänger for Cook," wrote *Fortune* reporter Adam Lashinsky in his book *Inside Apple*. "Tall, lean, and gray-haired, like Cook, Williams was said by Apple executives to look so much like his boss that from behind they could be mistaken for each other."

Philosophically, Williams seems a good match for Cook as well. Both are fitness freaks with a penchant for cycling, and both are intensely private about the lives they lead outside of Apple. Williams is frugal. For years, he drove a battered old Toyota with a busted passenger-side door, even after being elevated to a managerial role inside Apple, a position that came with millions of dollars in valuable stock options. With employees, he is said to be direct but fair, telling people what they need to do to solve a problem, without beating them up over it. "With Jeff, what you see is what you get," said Gerald Hawkins, a friend of Williams's and director emeritus of the Caldwell Fellows program at North Carolina State University. "And if he said he was going to do something, well, he'd do it."

Cook played a key role in training Williams for his job and has been effusive when describing him. When Williams was promoted to COO, Cook issued a gushing statement in which he said, "Jeff is hands-down the best operations executive I've ever worked with." Like Cook, Williams maintains a low profile, working very much behind the scenes at Apple. He wasn't mentioned once in Walter Isaacson's 2011 biography of Steve Jobs, and now he may be on track to be Apple's next CEO.

Since 2010, Williams has overseen Apple's entire supply chain, service and support, and social responsibility initiatives—the last being something that has grown in importance under Cook. One of Williams's triumphs on the iPod supply chain was reportedly establishing a deal that called for Apple to prepay suppliers like Hynix around $1.25 billion to secure flash memory for the iPod Nano. He also helped to speed up the iPod delivery process, making it possible for customers to buy an iPod online, have it custom engraved, and delivered within three working days. Touches like this are pure Cupertino magic—and in this case, it's Williams they have to thank for it. Williams is also said to be a key contact with supplier Foxconn.

"Jeff Williams is doing a phenomenal job," wrote Apple analyst Neil Cybart. "As senior vice president of Operations, Williams is tasked with making sure the Apple machine is well-oiled and in tip-top shape, not only capable of producing more than 100 million iOS devices in a quarter, but building flexibility into the system to handle annual hardware updates that would make most hardware companies quiver with fear. . . . He is executing at levels that few are able to achieve." He played an important role in the development of the first-generation iPhone, and has since led worldwide operations for both the iPhone and iPod. Today, he also supervises development for the Apple Watch.

Surprising Partnerships

To keep the company moving forward at top speed, Cook also led innovative partnerships with new companies. In May 2014, Apple announced the $3 billion acquisition of Beats Music and Beats Electronics. "Music is such an important part of all our lives and holds a special place within our hearts at Apple," Cook said in a press release. "That's why we have kept investing in music and are bringing together these extraordinary teams so we can continue to create the most innovative music products and services in the world."

During an interview with Charlie Rose in September 2014, Cook revealed the reason behind Apple's decision to buy Beats. Jimmy Iovine, a cofounder of the company, had earlier told him how great Beats Music was, he said, "so one night I'm sitting playing with [their service] versus some others, and all of a sudden it dawns on me that when I listen to theirs for a while, I feel completely different," Cook said. "And the reason is that they recognized that human curation was important in the subscription service." Human curation—the curation of tracks by human editors rather than algorithms—was mostly unique to Beats Music, with rival streaming services instead using computer algorithms to recommend music based on a listener's previous activity. "It's hard to describe, but you know when you feel it," Cook added. It was at that point he decided, "We need to do this. So that night—I couldn't sleep," he said. He put the acquisition plan in motion.

As part of the deal, which remains Apple's biggest acquisition to date, Beats cofounders Iovine and Dr. Dre joined Apple, and brought with them Beats president Luke Wood, marketing chief Bozoma Saint John, chief creative officers Trent Reznor and Ian Rogers, and chief operating officer Matthew Costello. "What Beats brings to Apple are guys with very rare skills," Cook later told Recode. "People like this aren't born every day. They're very rare. They get music deeply. So we get an infusion in Apple of some great talent."

It's unclear what roles Dr. Dre and Jimmy Iovine played after the Beats acquisition. Dre was largely absent, and Iovine gained notice mostly for his long and rambling introduction of Apple Music at a June 2015 event. But he was brought in for his music industry connections, and in that he may have been successful.

Bozoma Saint John became one of Apple's highest-profile executives. One of Apple's few female African American executives, she was a regular fixture at industry events and Apple's own product keynotes. She brought some much-needed glamour—as well as gender and color diversity—to Apple's top management. After several years at Apple, she left the company in June 2017 to become the chief brand officer at Uber. A year later, she jumped ship to become the chief marketing officer at Endeavor, a movie, sports, and fashion talent agency.

Before the acquisition, Apple had been selling Beats headphones and speakers directly through its online and retail stores, and it continued to offer the Beats Music streaming service until the unveiling of Apple Music just over a year later. Apple Music was built on the foundation laid by Beats, and continues to deliver the curated experience Beats subscribers fell in love with. It has also boosted its subscribership with exclusive partnerships with artists like Taylor Swift, Frank Ocean, Drake, and Chance the Rapper. Apple Music has become surprisingly successful, with thirty-eight million subscribers in March 2018, and was predicted to be on track to surpass Spotify in the United States by the summer of 2018.

Partnership with IBM–iOS in Enterprise

After updating the MacBook Pro with a Retina display in July 2014, Apple announced a surprising new partnership with IBM, a company Jobs had famously loathed since the early 1980s, when Apple and IBM first became huge rivals. Apple had always been a company that focused on ordinary consumers, and with the exception of schools and colleges, it had largely ignored business customers, especially the large global enterprises that IBM focused on. The partnership showed Cook's willingness to work with outside partners and to branch out into the business world. The alliance brought together the "market-leading strengths" of Apple and IBM with the promise of transforming enterprise mobility through a new class of more than a hundred industry-specific enterprise solutions. "iPhone and iPad are the best mobile devices in the world and have transformed the way people work with over 98 percent of the Fortune 500 and over 92 percent of the Global 500 using iOS devices in their business today," Cook highlighted in Apple's press release. "This is a radical step for enterprise and something that only Apple and IBM can deliver."

Cook explained that IBM was in the process of designing many different apps, for many different industries—including banking, manufacturing, and aerospace—the first ten of which would launch before the end of 2014. "This is an area where I think everybody's gonna win. We're gonna win, IBM's gonna win, and more importantly than us, the customer's gonna win," he explained to Charlie Rose. And indeed Apple and IBM did win, at least in market reaction. Apple shares climbed 2.59 percent in premarket trading the day the deal was announced, while IBM stock rose almost 2 percent. Shares in BlackBerry, which at the time was fighting to win back market share with its own enterprise solutions, tumbled nearly 10 percent on both the Nasdaq and Toronto stock exchanges.

The odd-couple pairing appears to have been successful. In July 2017, on the three-year anniversary of the partnership, the two companies said that more than a hundred enterprise iOS apps had been created across fifteen industries. In health care alone, Apple and IBM have developed a half dozen hospital-grade enterprise programs that are publicly available, and dozens of custom apps for specific customers, the companies said. More than thirty-eight hundred organizations, including hospitals and health systems, are using MobileFirst for iOS apps, according to Sue Miller-Sylvia, vice president of IBM MobileFirst for iOS Solutions.

The partnership was a big win for Cook in the age of BYOD (Bring Your Own Device), where workers often prefer to use their own devices for work. Twenty years ago, companies like Microsoft and Dell ruled the corporate world, selling huge volumes of computers to companies big and small. But in the age of BYOD, companies must adapt. Those that accommodate the BYOD trend are the ones that are winning, and Cook is leading the charge with successful big-business partnerships like the team-up with IBM.

iPhone 6 and Apple Pay

The unveiling of the iPhone 6 and iPhone 6 Plus, Apple's first large-screen phone, was described by Cook as "the biggest advancement ever in iPhone history." The radically redesigned handsets, with significantly larger Retina HD displays, faster LTE connectivity, and Apple's latest A8 chipset, quickly became the most successful Apple products released under Cook up to that point, with four million units sold in the first twenty-four hours, and more than ten million sold over launch weekend alone. The upgrades also brought Apple Pay, giving iPhone users for the first time the ability to make payments by tapping their device on a compatible NFC terminal.

Both the iPhone 6 and iPhone 6 Plus were greeted by overwhelmingly positive reviews, with many publications labeling them the best smartphones money could buy. They were praised for their performance, design, improved cameras, and their sharper and more colorful displays. But it wasn't long before the iPhone 6 and iPhone 6 Plus were making headlines for all the wrong reasons. It started with "Bendgate."

Users discovered that the new phones had a tendency to bend, particularly when carried in tight pockets. A YouTube video published by Lewis Hilsenteger of Unbox Therapy, which racked up tens of millions of views in just a few days, revealed that Apple's new aluminum unibody could easily be warped when pressure was applied to the back of the phone. The iPhone 6 Plus, with its substantially larger frame, was more susceptible to the problem than its smaller sibling. Apple denied that it was a widespread issue and said that just nine of the millions of units sold in the six days following the launch had been returned after bending. With their precision-engineered unibody enclosure, made of anodized aluminum, and inserts composed of stainless steel and titanium, the new iPhones met or exceeded Apple's "high quality standard to endure everyday, real life use."

Apple encouraged consumers to contact Apple Support if they experienced the "extremely rare" issue, and later invited journalists into its labs to demonstrate the durability tests its iPhone 6 and iPhone 6 Plus went through during product development. Support staff pointed worried customers with deformed devices toward the Genius Bar, where they could undergo a "Visual Mechanical Inspection." Assuming no abuse was identified by a Genius, and the damage was "within the guidelines" Apple had set, customers would be offered a free replacement.

Embarrassing iOS 8.0.1 Bug

Before Bendgate had a chance to blow over, Apple frustrated iPhone 6 owners with a software update, iOS 8.0.1, that contained a bug preventing around forty thousand users from making calls. The handsets were just weeks old at this point, and those affected were unable to connect to a cellular network. Another bug played havoc with the Touch ID sensor used for fingerprint recognition. "We are actively investigating these reports and will provide information as quickly as we can," read an Apple statement issued on September 24. "In the meantime we have pulled back the iOS 8.0.1 update." Those who hadn't already installed the release no longer had access to it. Those with an iPhone 6 who had updated early were forced to suffer its glitches until iOS 8.0.2 rolled out on September 26, just two days after 8.0.1 was released.

Despite some teething troubles, the iPhone 6 is still considered one of Apple's most notable refreshes. It signaled a major change in the company's approach to smartphones—and particularly smartphone displays—which it previously believed should be small enough to operate with just one hand. It wasn't until Apple started losing customers to larger Android devices that it finally listened to calls for bigger screens. Data from Kantar Worldpanel ComTech revealed that 26 percent of people buying a Samsung smartphone in the three months leading up to May 2014 were switching from the iPhone, compared to just 12 percent the year before. In many of Apple's key markets around the world, including the United States, the iPhone's market share had been falling as Android's climbed.

But the introduction of a bigger screen with the iPhone 6 series soon changed that. Combined, the new handsets enjoyed significantly faster adoption rates than the iPhone 5 and iPhone 5S, and after less than a year on the market, they had already become the most popular iPhones in use, accounting for 40 percent of Apple's overall market share. They even helped Apple attract "a higher rate of switchers [from Android] than we've seen in previous iPhone cycles," Cook revealed during a call with investors in April 2015.

Apple Pay

During the iPhone 6 keynote, Cook also introduced the world to Apple Pay and the company's wildly ambitious plan to replace credit cards and cash. Shoppers in the United States alone were spending $12 billion using credit and debit cards every day at this point, and Apple wanted a piece of that pie. Rivals like Google had already begun their efforts to standardize mobile payments, with little success, but Cook was confident that Apple's solution would be a winner, partly by focusing on privacy.

"Most people who have worked on this have started by creating a business model that was centered around their interests instead of focusing on the user experience," Cook explained on a stage in front of the keynote audience. "We love this kind of problem. This is exactly what Apple does best." And unlike the others, Apple had no intention of using its payments service to turn customers into products by mining their data.

"You are not our product," Cook insisted during an appearance at the Goldman Sachs Technology and Internet Conference in February 2015.

"There's no reason why anyone . . . need[s] to know where you're buying something, what you're buying, how much you're paying. . . . I don't want to know any of that. It's none of my business, frankly." Cook was also keen to highlight Apple Pay's advantages in security, and how users would never have to worry about their credit card information ending up in the wrong hands.

Cook believed that "Apple Pay will forever change the way all of us buy things," and fans couldn't wait to try it out. During an appearance at a *Wall Street Journal* technology conference a week after the service was unveiled, Cook revealed that one million cards had been registered with Apple Pay during its first seventy-two hours of availability. This figure, which he keenly noted was "more than the total of all the other guys," meant Apple's mobile payments service was already the leader in "contactless" payments. Early adopters were impressed with Apple Pay's focus on ease of use. "I'm getting flooded with emails from customers," Cook said. "It's sort of that 'ahh' moment—you use the phone and it's all you have to do."

By January 2015, more than 750 banks and credit unions had added Apple Pay support, and more than two out of every three dollars spent using contactless payment platforms were going through Apple Pay. "I'm unbelievably shocked, positively shocked, at how many merchants were able to implement Apple Pay during the holiday," Cook said during an Apple earnings call. He proclaimed, "2015 will be the year of Apple Pay."

To date, Apple has never revealed exactly how many people have signed up for Apple Pay, or how many people use the service on a regular basis, but the company occasionally discloses impressive figures and milestones. In June 2016, it confirmed to *Fortune* that a million people were signing up for Apple Pay every week—five times the amount the service was attracting a year earlier. In May 2017, during Apple's second quarter earnings call, Cook boasted about Apple Pay's impressive growth after the service saw a 450 percent increase in transactions over the same three-month period a year earlier.

By this time, Apple Pay had expanded its reach to fifteen territories throughout North America, Europe, Australia, and Asia—but its adoption wasn't quite living up to Cook's expectations. He admitted to shareholders in February 2018 that Apple Pay had "taken off slower that I personally would have thought if you asked me sitting here a few years ago." Nonetheless, Apple's CEO remains optimistic about the future of Apple Pay and similar mobile payments services.

At a 2018 Apple shareholder meeting, Cook was optimistic about the future of Apple Pay and other contactless payment systems. "I'm hoping to be alive to see the elimination of money," he said.

Apple Pay seems to be gathering momentum. During Apple's quarterly earnings call in July 2018, Cook told analysts and investors that Apple Pay had been used for "well over a billion transactions" during the third fiscal quarter of 2018, which is a big number, and far outstrips the competition. Cook also revealed plans for the service to expand into CVS and 7-Eleven stores throughout the United States, and to Germany later in the year. Apple Pay may be off to a slower start than Cook would have liked, but his contactless payment dream seems to be gathering steam.

Cook's First Major Product: Apple Watch

On a bright sunny day in September 2014, Cook finally unveiled Apple's much-anticipated Apple Watch, which he called "the next chapter in Apple's story." It featured a heart rate monitor and fitness tracking capabilities that gave Apple the opportunity to market the device to health fanatics who wanted to monitor and track their workouts. Its built-in Activity app encouraged users to stand, exercise, and keep fit—and offered virtual achievement medals for those who reached their daily goals. It was obviously a device very close to Cook's fitness-fanatic heart. He described Apple Watch as "a precise timepiece, a new intimate way to communicate from your wrist, and a comprehensive health and fitness device."

Fans would have to wait until April 2015 to get their hands on Apple's first wearable. Apple Watch wasn't quite ready to start shipping, and Apple wanted to give developers time to build apps for it, Cook explained. This would ensure that early buyers could download their favorite apps the day they received their device. Facebook and Twitter were two of the largest companies already working on apps for watchOS, the name Apple gave to its wearable operating system, a custom version of iOS. Apple received "overwhelmingly positive" feedback from early adopters after Apple Watch finally went on sale, Cook said, but to date, the company refuses to reveal exact sales figures for the device. Based on third-party estimates, Apple Watch easily outsells rival smartwatches from the likes of Samsung, Motorola, and LG. It is believed that it took less than twenty-four hours for the original model to surpass the total number of Android Wear devices sold up to that point, which was around 720,000.

In an interview with the high-end watch website Hoodinkee, Jony Ive revealed that the Apple Watch was the first major product with no input from Steve Jobs. "We didn't talk about watches, or us making a watch," Ive said, referring to discussions with Jobs. "I don't remember him wearing one either."

The executive team started brainstorming a few weeks after his death, and in fact, the idea for the watch came indirectly from Jobs's demise. "The first discussion took place in early 2012, a few months after Steve's passing," Ive said. He added, "It caused us to take time, pausing to think about where we wanted to go, what trajectory we were on as a company, and what motivated us." The answer was the Apple Watch.

To consolidate his role as Apple's new leader in 2013 and 2014, Cook explored opportunities in new markets, sought out interesting partnerships, and peddled ruthless innovation of the iPhone and development of the Apple Watch. By the end of November 2014, after its stock price hit a record high, Apple's market capitalization surpassed a staggering $700 billion for the first time. At the time, the iPhone maker was worth twice as much as Google, and $300 billion more than ExxonMobil, the second most valuable company on the planet. Any doubts about Cook's ability to fill Jobs's shoes and lead Apple to even greater success were being well and truly stamped into the ground.

Chapter 8

A Greener Apple

Apple is now considered one of the greenest companies in the technology industry, but it wasn't until Cook was permanently installed as CEO that its environmental efforts became entirely genuine.

"Sustainability was pretty much an afterthought at Apple," said Abraham Farag, a former senior mechanical engineer of product design, who worked at Apple between 1999 and 2005. Farag labeled the company's environmental efforts as little more than "lip service," conducted only to appease the activists and the consumers who cared. At the time, Apple had just one person dedicated to examining environmental impact, and there was "no way one person could have much impact without very strong top-down support" from Apple's executive team, "which she did not get," Farag continued. "She certainly tried, but it was an impossible task."

Farag recalls a time when sustainability organizations were pushing for recycling codes to be stamped on products that weighed more than 25 grams, to help with the recycling process, but Apple refused because the codes were not "pretty." There was "no way we could alter the design for

141

that consideration," he explained. "Pure looks trumped any responsible consideration for sustainability." With Jobs at the helm, Apple seemed to be doing the bare minimum to be seen as environmentally friendly, but reducing the company's footprint was never a priority. At this time, in the mid-2000s, Cook was senior vice president for worldwide operations, and not in charge of these policies.

Under Steve Jobs, Apple made a series of environmentally unfriendly decisions, which were reflected in many Greenpeace reports throughout the 2000s that lambasted Apple for its destructive impact on the environment. Greenpeace's first critical report, published in 2007 and titled *Missed Call: iPhone's Hazardous Chemicals*, detailed the discovery of several damaging materials used in components for the original iPhone. The analysis found evidence of several hazardous substances regulated by the European Union, including antimony, bromine, chromium, and lead. It also found evidence of the plastic polyvinyl chloride (PVC), which Apple had pledged to remove from its products. "The fact that a product brought newly to the U.S. market in June 2007 still utilizes PVC and brominated flame retardants suggests that Apple is not making early progress towards its 2008 commitment to phase-out all uses of these materials, even in entirely new product lines," the report concluded. "If Apple really wants to reinvent the phone, it needs to design out all hazardous substances and materials from its handsets and peripherals."

Following the Greenpeace berating, Jobs leapt to Apple's defense, publishing a letter titled "A Greener Apple," in which he highlighted a number of things the company had done, and was still doing, to make its products less hazardous. This included eliminating CRT displays from its products entirely in 2006, and phasing out PVC and BFRs in all products by the end of 2008. He also pointed out that Apple was operating recycling programs in countries where more than 82 percent of Macs and iPods were sold. "Apple is ahead of, or will soon be ahead of, most of its competitors in these areas," he insisted. Soon afterward, Apple started publishing annual environmental responsibility reports, a practice it continues today, detailing the company's green initiatives.

In response to Jobs's letter, Greenpeace publicly commended Apple for its efforts. They were particularly impressed by Apple's promise to eliminate PVC and BFRs and encouraged rival computer makers—Acer, Dell, HP, Toshiba, and Lenovo—to follow in Apple's footsteps: "There's no excuse for any of these companies not to follow Apple's lead on toxic chemicals elimination now and not wait."

The iPhone 3GS, launched in 2009, was one of the first products that Apple claimed was free of PVC and BFRs. But Apple would remain a focus for Greenpeace in the years that followed, particularly as interest in the company grew because of the iPhone and iPad. In their quarterly *Guide to Greener Electronics* published in June 2009, Greenpeace criticized Apple for "unreasonably high threshold limits for BFRs and PVC in products that are allegedly PVC-/BFR-free." It's not clear whether this was because Apple didn't remove the chemicals it said it had, or if its contractors hadn't. (Today, under Cook's guidance, Apple exhaustively audits all the components and subcomponents of its products to ensure they meet strict specifications, including their chemical makeup.) Greenpeace also gave the company low marks for its measly efforts to use recycled plastics and renewable energy. But later that year, Apple climbed several places in Greenpeace's ranking in December, as it had committed to making all products free from PVC, except for cables. It was also awarded more points for an increase in use of renewable energy. But Greenpeace soon become the least of Apple's environmental concerns.

Pollution and Poison

More severe criticism of Apple's practices came from China, where anti-pollution activists were accusing the company of secretly degrading the environment and poisoning supply chain workers. A January 2011 study, carried out by a coalition of China's leading environmental groups, ranked Apple at the bottom, along with twenty-nine other major technology firms. The report was published just a few months after it was reported that Chinese workers had been hospitalized following exposure to the chemical n-hexane, which was used to clean iPhone screens.

"Behind their stylish image, Apple products have a side many do not know about—pollution and poison," read a report from the Green Choice Initiative. "This side is hidden deep within the company's secretive supply chain." An accompanying video, which cited some of the sixty-two people who had been hospitalized in Taiwan, demanded an explanation from Jobs. "We want to ask you whether or not you should be responsible for the supplier companies you have chosen?" they asked. "When you look down at the Apple phone you are using in your hand and you swipe it with your finger is it possible that you can feel as if it is no longer a beautiful screen to show off, but the life and blood of us employees and victims?"

The poisonings happened at a Wintek factory in Suzhou, where managers reportedly switched to n-hexane—a chemical that has the potential to cause nerve damage for up to two years—because it dried more quickly than alcohol, and therefore increased efficiency. At the time, Apple would not confirm or deny that it had a relationship with Wintek, despite repeated requests for comment from Green Choice over several months. (At the time, Apple was super secretive about its suppliers. The company routinely refused to identify who it worked with. These days, under Cook, it is much more open and publishes a comprehensive list of all its suppliers, updated annually.)

"This attitude means it is impossible to have any public supervision over their supply chain," said Chinese environmentalist Ma Jun, director of the Institute of Public and Environmental Affairs. "When environmental violations become public knowledge, they should not use commercial confidentiality as an excuse for silence. This is different from other leading brands." Nokia and Motorola had already responded to questions about their involvement with Wintek, but Apple said it would not comment on individual allegations.

Ma, a former investigative journalist, had published a report earlier in 2011 titled *The Other Side of Apple*, in which he detailed the harrowing effects that working in Apple's supply chain had on some of the workers. In an accompanying video, sick workers described their symptoms among clips of Jobs showcasing new Apple products. Despite making a significant impact, the report drew little attention from Apple, but five months later, Ma published *The Other Side of Apple II*, detailing another ten case studies of environmental violations at suspected Apple suppliers.

Ma and the Institute of Public and Environmental Affairs accused Apple's suppliers of causing severe pollution that endangered the public's health and safety. In a forty-six-page report, published in September, following visits to many of the factories throughout China, the group said that they suspected "dozens" of Apple suppliers were discharging polluted waste and toxic metals into surrounding communities, while twenty-seven had been found to have caused environmental problems.

Initially, Apple released a statement reflecting the ongoing initiatives already in place for its suppliers. "Apple is committed to driving the highest standards of social responsibility throughout our supply chain," said

Steve Dowling, a company spokesperson, in response to the report. "We require that our suppliers provide safe working conditions, treat workers with dignity and respect, and use environmentally responsible manufacturing processes wherever Apple products are made." Dowling also insisted that Apple had been aggressively monitoring factories in its supply chain and carrying out regular audits.

But the company also acknowledged Ma's report, requesting a phone call to "build a relationship with Ma rather than try to control him," according to Yukari Kane's *Haunted Empire*. "So began delicate conversations between Ma and Apple." Progress was slow initially, and early meetings proved uneasy and uncomfortable, Ma told Kane. But soon after Jobs's passing, "there was a breakthrough." The breakthrough was attributed to Cook, who would later devote more of the company's resources to supplier responsibility.

Change for the Better

Cook has made his love of the environment clear in initiatives he has taken at Apple. Since he became CEO of the company, he has transformed Apple's environmental policies and approach to sustainability. Just a few weeks after he became CEO, one of his top lieutenants in operations, Bill Frederick, reached out to Linda Greer, an environmental toxicologist with the Natural Resources Defense Council, a Washington, D.C.–based nonprofit that works with corporations to help improve their environmental impact.

Frederick admitted that environmental safety had not been a priority for Apple under Jobs, but he assured the NRDC that company executives were now considering ways in which they could keep better tabs on suppliers. Apple requested a meeting and the NRDC agreed—with the proviso that Ma, who had already started working with the NRDC, could attend.

"The meeting lasted for five hours," writes Kane. "Some of the executives were defensive at times, and Apple continued to withhold details such as the discrepancies that they found between their audit and Ma's investigation. But they also acknowledged that they needed to be more transparent. At the end, the company agreed to do its own research to confirm Ma's findings and then to address whatever problems they found." Greer found that Apple was doing more than most of its rivals to make amends, and Ma felt that the company was "changing for the better" less than a month after Cook took the reins. Whereas Jobs was—at best—indifferent to the environmental impact of Apple's supply chain, Cook was determined to tackle the subject head-on. Apple's approach to the environment had taken a 180-degree turn.

Dirty Data

Cook's focus on improving Apple's environmental efforts was evident, but fixing the existing mess would take time and effort. Apple's emissions per product were at an all-time high when Cook was appointed CEO, and as the company continued its push toward cloud-based services, the Greenpeace battering continued. Apple was labeled the "least green" tech company in a 2011 report on the environmental impact of cloud computing, which asked, *How Dirty Is Your Data?* Cloud computing was becoming a major industry force, with every tech company from Amazon to Zendesk investing in giant, power-hungry data centers to run their online services, and Apple was at the top of the pack.

Apple received the lowest percentage on the "Clean Energy Index," and the highest for "Coal Intensity," after choosing to locate its most recent iCloud data center in North Carolina, a place with "an electrical grid among the dirtiest in the country," where coal-powered energy could be acquired cheaply. "The fact that the alternative location for Apple's iData-Center was Virginia, where electricity also comes from very dirty sources, is an indication that, in addition to tax incentives, access to inexpensive energy, regardless of its source, is a key driver in Apple's site selection," read the Greenpeace report.

With no response from Apple as the months went by, activists started protesting outside Apple Stores around the world. Greenpeace activists in New York City released black helium-filled balloons inside the company's flagship Fifth Avenue store that blocked the sun from shining through its glass ceiling. May 2012 saw two Greenpeace activists barricade themselves in a giant iPod, retrofitted from an eight-foot-tall, ten-foot-wide survival device previously used in protests to prevent Arctic drilling, in front of Apple's Cupertino headquarters. The duo played audio messages from fans around the world, who pleaded with Apple to swap coal for renewable energy, to employees who passed by the entrance to the building. Four more activists, dressed as iPhones, later turned up to support them, with TVs strapped to their torsos displaying messages from supporters on social media.

"Apple's executives have thus far ignored the hundreds of thousands of people asking them to use their influence for good by building a cloud powered by renewable energy," said Greenpeace USA executive director Phil Radford in response to the protest. "As Apple's customers, we love our iPhones and iPads, but we don't want to use an iCloud fueled by the smog of dirty coal pollution."

Cook Sets to Work

The protests would probably have been ignored under Jobs's leadership, but with a growing conscience under Cook, Apple acted fast. In a statement published just two days later, the company promised a switch to using renewable energy for its Maiden, North Carolina, data center by the end of 2012. Apple had already started building a one-hundred-acre solar array and a biogas energy plant on the site, but the company was still buying a large chunk of energy from the coal-powered Duke Energy. Apple vowed to source that energy from local renewable energy providers instead, and promised that another of its data centers, in Newark, California, would switch to renewable energy in early 2013.

Greenpeace released a statement in response, praising Apple's efforts. "Apple's announcement today is a great sign that Apple is taking seriously the hundreds of thousands of its customers who have asked for an iCloud powered by clean energy, not dirty coal." But still, Greenpeace insisted it would not rest until Apple, and other major technology firms, would commit to using renewable energy as they build new data centers going forward. "Only then will customers have confidence that the iCloud will continue to get cleaner as it grows."

Apple's China crisis would prove difficult to extinguish, too. In February 2013, one Apple supplier, RiTeng Computer Accessory, part of the iPad supply chain, was penalized by Chinese environmental authorities for dumping so much waste into a local river that it turned "milky white." Residents raised concerns after noticing the color of the water, and a subsequent investigation found that it contained cutting fluid and oil, which was traced back to storm drains owned by RiTeng. Only eighteen months before, an explosion in one of the company's buildings had sent sixty-one workers to the hospital for monitoring. Things were not looking good for Apple.

Enter the EPA

Cook, fighting an uphill battle and recognizing that Apple needed help, expanded the company's executive team with the 2013 appointment of

Lisa Jackson, the former head of the U.S. Environmental Protection Agency (EPA) for four years during the Obama administration. The appointment of Jackson signaled that Cook was serious about cleaning up Apple's act.

Hiring Jackson was the first step in Cook's larger plan to make Apple a greener company. "Jackson can make Apple the top environmental leader in the tech sector by helping the company use its influence to push electric utilities and governments to provide the clean energy that both Apple and America need right now," Cook said, announcing the appointment at the D11 technology conference in May 2013.

In an interview at Apple Park in March 2018, Jackson reinforced Cook's new mission. "Tim obviously is very interested in environmental issues," she said. "He puts a great value on the environment." Cook has a deep love of the outdoors, spending most of his spare time hiking and biking in California's Yosemite National Park. "That's part of who he is," Jackson said. She, too, is determined to preserve the environment. "It's all about nature," she said. "You can't get it back, once we destroy it. And it gives us so much, from peace of mind to clean water, to clean air. Those are all things that in varying parts, we cannot succeed as humans without."

Jackson is an engaging and gregarious African American woman in her mid-fifties, who grew up around the same time and not far from Cook in New Orleans. She joined the EPA as a staff-level engineer a few years after college and rose quickly through the ranks. In 2008, she was tapped by the Obama administration to be administrator of the EPA, overseeing about seventeen thousand employees as the agency's first African American administrator. Though she earned a reputation as hardworking and a consensus builder, she was accused of having close ties to industry and quit in 2012 after a series of political headaches. This led her to Apple less than a year later.

On her first day of work at Apple in May 2013, Cook had much to learn from her. He asked her, "What is Apple doing wrong? What is Apple doing right? What can we do better? Help us outline a vision and then help us get there." He was clearly invested from the start. Bringing Jackson on was the first step, but Cook was eager to get down to business. He was determined to take an active role in the company's environmental initiatives, as his predecessor had not.

The mission Cook wanted to take on, he told Jackson, was to "leave the world a better place than we found it," a statement that's oft been repeated in Apple's environmental reports, promotional videos, and inspirational posters. As one of the world's most powerful companies, Apple could use its massive resources to enact real change. Cook's "big vision was, 'Apple's a big company . . . how can we use Apple as a force for good? How can we make sure that a company of our size and scope is . . . attacking the big issues before us?'" He ultimately decided to focus on three areas: addressing climate change, using greener materials in products, and helping safeguard the planet's resources.

Greenpeace's Gary Cook praised Tim Cook's leadership on his commitment to using only renewable energies and materials. He said, "Tim Cook thinks it's important. He's been quite outspoken on the need for action on climate change, on Apple's responsibility to address climate change. I think him going off and hiring Lisa Jackson, a former EPA administrator, is an indication that he thought this was a top priority for Apple and something that needed to be addressed long-term." Getting Greenpeace's approval was a great first step in Cook's plan to change Apple's environmental policies for the better. But Cook wouldn't stop there.

A Force for Good

Apple's efforts to be green were now much larger than they had ever been in the company's thirty-seven-year history. It was clear that environmental issues were high on Cook's agenda and no longer just an afterthought to appease the activists. Apple had become "a force for good in the world beyond our products," Cook said in October 2013 in an earnings call with analysts and investors. "Whether it's improving working conditions or the environment, standing up for human rights, helping eliminate AIDS, or reinventing education, Apple is making substantial contributions to society."

Due to Tim Cook's new mission, Apple quickly started climbing Greenpeace rankings, reduced emissions per product year after year, and increasingly relied on renewable energy, rather than coal, to power manufacturing facilities, offices, and retail stores around the world. In 2014, Apple was named one of the cleanest data center operators in the world in one Greenpeace report. Apple had come a long way—Tim Cook was changing the company and making the world a better place.

And Cook wanted to make sure Apple was a leader in environmental conservation. If Apple made the environment a priority, other companies would follow. Apple needed to be "one of the pebbles in the pond that creates the ripple," Cook said during an appearance at Climate Week NYC in September 2014, using a phrase from Robert F. Kennedy. Apple would not accept "a trade-off between the economy and the environment," he told UN Framework Convention on Climate Change executive secretary Christiana Figueres. "If you innovate and you set the bar high, you'll find a way to do both, and . . . you must do both, because the long-term consequences of not addressing the environment are huge."

By 2014, Apple managed to remove PVC entirely from its products—

and it took more than six years to find a good replacement substance. Lisa Jackson said it was important for Apple to remove PVC from power cords because they knew that people around the world "were burning cords to get to the copper inside. That's a really dangerous situation from a health perspective. It's also not great in the manufacturing process. And we were hoping that in doing that others would follow suit. But it hasn't come to be. So you have to do what's right. Tim says this all the time. You have to do what's right, because it's right." This sums up Apple's environmental policies under Cook. He hasn't been making these changes for show—he truly cares about doing the right thing.

Cook Ramps Up Solar

Another way Cook addressed Apple's responsibility to the environment was by implementing renewable energy initiatives. In February 2015, Apple announced it was partnering with First Solar to build an $850 million solar energy farm in California. Located in Monterey County, the farm, which would become Apple's fourth upon its completion in 2016, would produce enough energy to power nearly sixty thousand homes. Apple and First Solar signed a twenty-five-year power purchase agreement, then considered the largest commercial power agreement in the industry, which would provide Apple with 130 megawatts of power from the project.

"Apple is leading the way in addressing climate change by showing how large companies can serve their operations with 100 percent clean, renewable energy," said Joe Kishkill, First Solar's chief commercial officer, in a statement confirming the deal. "Apple's commitment was instrumental in making this project possible and will significantly increase the supply of solar power in California. Over time, the renewable energy from California Flats"—which was still under construction in July 2018—"will provide cost savings over alternative sources of energy as well as substantially lower environmental impact."

In response to the solar deal, Cook acknowledged that climate change was a real threat and emphasized that "the time for talk is passed [sic]. The time for action is now." And Greenpeace took note: "Apple still has work to do to reduce its environmental footprint, but other Fortune 500 CEOs would be well served to make a study of Tim Cook," read a statement from the group after the deal was announced.

In October 2015, Apple announced plans to build 200 megawatts of solar projects—enough to power 265,000 homes—in China as part of its effort to push local suppliers to be more sustainable. The project would "begin to offset the energy used in Apple's supply chain," which was still responsible for more than 70 percent of Apple's carbon footprint. Cook said, "Climate change is one of the great challenges of our time, and the time for action is now. The transition to a new green economy requires innovation, ambition and purpose. We believe passionately in leaving the world better than we found it and hope that many other suppliers, partners and other companies join us in this important effort."

Foxconn, one of Apple's largest manufacturing partners, was already in support of this push, with plans to build 400 megawatts of solar projects in China's Henan Province by 2018. The company was committed to generating as much clean energy as its Zhengzhou factory typically consumed during final production of the iPhone. This was a huge deal, considering all the issues Apple and Foxconn have had over the years with respect not only to worker conditions but also to the environmental impact of the factories.

In an interview with the *Washington Post* to mark his fifth year as Apple CEO, Cook said he was incredibly proud of how Apple had "stepped up our social responsibility," and he highlighted this as one of the ways in which the company had evolved under his leadership. "We've had environmental work going on at Apple for decades, but we didn't talk about it, and we didn't set aspirational kind of objectives," he explained. "We used the same philosophy we do with our products, which is you unveil them when you've finished. But we stepped back and re-evaluated that and said, 'You know, if we wait until you do that, we're not helping anyone else get there.'"

Then, in December 2016, Apple made another significant sustainability deal with a Chinese renewable energy firm, Goldwind Science & Technology, the world's largest wind turbine maker, as part of Cook's plan to shift the company's entire supply chain to renewable energy. It was Apple's first foray into wind power and its largest clean energy project to date, according to Jackson. As part of the agreement, Apple would obtain 30 percent equity in four companies—all owned by Beijing Tianrun New Energy, a subsidiary of Goldwind—running wind power projects in the provinces of Henan, Shandong, Shanxi, and Yunnan that would produce 285 megawatts of wind power for local suppliers. The deal appeared to be in trouble in July 2017 when Goldwind reported a contract dispute with Apple that had affected the company's profitability, and now it's unclear if the deal is still active. But a year later, in July 2018, Apple joined ten of its suppliers in founding a $300 million fund to help its Chinese supply chain transition to green energy. The China Clean Energy Fund will invest and help develop more than 1 gigawatt of green energy in China (enough to power one million homes) and then connect that to its supply partners.

100 Percent Renewable

All of Apple's renewable energy deals paid off. On Earth Day in April 2018, Apple announced that its facilities around the world were running on 100 percent renewable power. That included all the data centers, retail stores, and offices around the world, like the giant new Apple Park campus.

Jackson was at pains to point out that Apple wasn't pulling any tricks, like buying carbon offsets. The company had invested in new sources of renewable energy—like its giant solar farms and solar panels on the roofs of many buildings—to ensure that it wasn't monopolizing the current supply. "We're really adamant that if we're in a location buying clean power, we don't want to buy up all the power," she said. "Because then somebody comes along and they can't get clean power because Apple or an Apple supplier has just bought it all. So a lot of that represents new additional clean power on the grid."

Apple's facilities, however, are just a small part of the company's total carbon footprint (about 27.5 million metric tons in 2017). Up to 77 percent of the footprint is in the supply chain, and Apple is adamant about reducing this as well. Jackson said Apple has outlined a very aggressive goal of adding 4,000 megawatts of new clean energy in partnership with its supply chain by 2020. These will cover about a third of the current supply chain; the remaining two-thirds still rely on dirty power. Jackson said it will take about four years to get a third of the supply chain running on renewables, and she hopes the remaining two-thirds will be running on renewable energy in a similar time frame, about eight years. "I would expect us to continue on that kind of trajectory," she said. They are off to a good start. So far, fourteen suppliers have committed to going 100 percent renewable for their Apple operations.

Apple is unique in starting sustainability initiatives in its supply chain. Greenpeace's Gary Cook noted that "they are the only company who have extended their 100 percent renewable commitment to also include their suppliers. It's starting to change conversations in the sector about what does corporate responsibility look like, or climate responsibility look like for a company like Apple." He said it's still early days, but he expects it to have a domino effect on other companies.

Companies like HP and Ikea have started to make commitments about their own supply chains. "It may take a few years but I think companies like Samsung, if they fail to address their carbon footprint and do not make a transition to renewable energy, they could find themselves at a real competitive disadvantage as more and more companies become much more demanding of their suppliers and have higher expectations about their environmental performance." Samsung, which actually makes most of its revenue not from refrigerators or televisions but from selling components like chips and screens to other companies like Apple, has only 1 percent of its global operations running on renewable energy, according to Gary Cook. In July 2018, several months after I spoke to him, Samsung announced that it would transition all of its operations in the United States, Europe, and China to 100 percent renewable energy by 2020.

Gary Cook added that Apple, Facebook, and Google are doing much better than Amazon and the big Chinese online companies—Baidu, Alibaba, and Tencent—which haven't done much to make their operations green yet. "Amazon is just going for broke," he said. "They've actually deployed a lot of renewable energy but it's not enough." Amazon's rapid growth is outpacing the supply of renewable energy. Apple under Tim Cook appears to be keeping up the company's emerging as it continues to be a leader in sustainable supply-chain practices, and hopefully more companies will continue to follow its lead.

Closed-Loop Supply Chain

In April 2017, Apple attracted further praise from Greenpeace with plans for a closed-loop supply chain, which would one day see its iPhone and other devices made entirely from recycled materials. Cook's intention was to change the very nature of the supply chain. The goal is to "one day stop mining from the earth altogether," explained Apple's 2017 environmental responsibility report. "It sounds crazy, but we're working on it. . . . One day we'd like to be able to build new products with just recycled materials, including your old products. It's an experiment in recycling technology that's teaching us a lot, and we hope this kind of thinking will inspire others in our industry."

March 2016 saw Apple take a big step toward its goal for a closed-loop supply chain with the introduction of Liam, a robot that pulls apart the iPhone 6, the company's most successful smartphone to date, so that all of its parts can be recycled. There are currently two Liam robots in operation— one in the United States and one in the Netherlands—each of which has the ability to disassemble a complete phone every eleven seconds, 1.2 million units every year.

"Traditional e-waste recycling can only recover a handful of the materials actually used in today's electronics," explains Apple's guide to the iPhone recycling robot. "Liam lets Apple address this problem by producing eight different material streams that can be sent for targeted material recovery. As a result, end-processors can recover a more diverse set of materials at higher yields than ever possible before." Apple calls Liam, which is still considered a research and development project, "a critical step in the journey toward establishing a closed-loop supply chain for Apple. It is also a vehicle to drive innovation in the recycling industry." In April 2018, a few days before Earth Day, Apple debuted a new version of Liam that can disassemble nine versions of the iPhone instead of one, and sort the parts to better recover recyclable materials that traditional recyclers can't. Called Daisy, the robot can process two hundred iPhones an hour and is located in Breda, Netherlands.

At present, Apple is experimenting with using recycled tin from soldering iPhones and aluminum recycled from iPhones in enclosures for the

Mac Mini. The advantage of using recycled aluminum from iPhones is that's it's very good quality, often better than can be sourced from raw material suppliers, because the quality of those materials can vary widely. Currently, the percentage of aluminum recycling is pretty small, about 4 percent. But Jackson said she expected Apple's recycling efforts to get better fast. She said that in 2018 Apple would open the tech industry's first material recovery lab that is "devoted entirely to getting material back out of products and using it again."

Apple also announced a breakthrough in finding a carbon-free way to smelt aluminum. In May 2018, Apple revealed that it had teamed up with Alcoa and Rio Tinto Aluminum to devise a method of smelting aluminum that eliminates direct greenhouse gas emissions. Apple called it a "revolutionary advancement in the manufacturing of one of the world's most widely used metals," which could reduce carbon dioxide production by 6.5 billion tons a year in Canada alone.

Aluminum has been produced the same way for over 130 years, using a process pioneered by Charles Hall, the founder of Alcoa Corporation, in 1886. The process involves applying a strong electrical current to aluminum oxide, its naturally occurring state, to remove the oxygen. This happens inside large smelters that are lined at the bottom with a carbon material that acts as an electrode and burns during the process, giving off carbon dioxide. It currently contributes to the 21 percent of greenhouse gas emissions that come from the industrial sector.

In 2015, three Apple engineers started searching for a cleaner way to mass-produce aluminum. They discovered that Alcoa had designed an entirely new process that replaces the carbon with an advanced conductive material that releases oxygen instead of carbon dioxide. But Alcoa needed a partner to move things forward. Apple's business development team brought in Rio Tinto, which has extensive expertise in smelting technology.

Alongside the Canadian government, as well as the aluminum smelting companies, Apple invested $144 million to fund a patented aluminum manufacturing process. Alcoa and Rio Tinto soon formed a joint venture they dubbed Elysis, which is committed to developing the new process for large-scale production and commercialization by 2024. "We are proud to be part of this ambitious new project, and look forward to one day being able to use aluminum produced without direct greenhouse gas emissions in the manufacturing of our products," Cook said in a statement.

Apple once again is a leader in developing new, sustainable technologies that will have a positive impact on our world. As of 2017, all of its final assembly sites around the world have been certified as contributing zero waste to landfill, and suppliers working with Apple have now introduced energy efficiency improvements that reduced upward of 320,000 annualized metric tons of greenhouse gas emissions in 2017 alone.

Sustainable Forests

Despite all the progress Apple has made in renewable energy and recycling, Greenpeace noted that there was still room for improvement in some areas, such as packaging sustainability and emissions reporting. In an April 2016 Greenpeace report on Apple's sourcing of sustainable paper and packaging, Gary Cook criticized Apple for appearing to value weak forestry standards from the Programme for the Endorsement of Forest Certification (PEFC), which includes, in the United States, the controversial Sustainable Forestry Initiative (SFI) label. According to Greenpeace, these PEFC and SFI standards are "greenwashing"—weak standards that contribute to controversial logging practices, forest destruction, and trampling on the rights of indigenous peoples and forest-dependent communities.

After this report, Apple made a commitment to source 100 percent of its packaging from sustainable resources, and as of 2018 it is 99 percent there, according to Lisa Jackson. Apple has partnered with groups in the United States and China to work toward the tougher Forest Stewardship Council Certification standard. In the United States, Apple is helping manage thirty-six thousand acres of sustainable "working forests" in Maine and North Carolina (in conjunction with the Conservation Fund), and in China, the company has three big paper "plantations" (equaling more than seven hundred thousand acres) in conjunction with the World Wildlife Fund.

In about two years, Jackson said, Apple has gotten to the point where these forests are producing enough sustainably managed wood to cover the company's needs. The company is also reducing how much virgin paper is needed for packaging, using more recycled paper, and making packaging smaller and lighter. Jackson said that Apple is trying to phase out plastics like styrofoam in all of its product packaging in favor of paper, a renewable resource.

Unfortunately, it's not all that easy. Apple is struggling to replace styrofoam in the packaging of big, bulky products like the iMac. "We came up with a paper solution but we weren't happy with it," Jackson said, "because we don't think you should sacrifice the customer experience and delight and expectation about quality and everything Apple means. . . . But we will get to a sustainable solution." Given her confidence, it should only be a matter of time before Apple's designers come up with something workable.

A Dedicated CEO

Thanks to Cook's dedicated approach, Apple is ahead of the curve on environmental issues. At "other companies . . . the CEO is not that involved," says Greenpeace's Gary Cook. "For Apple at least, they've really attached their brand to the environment in a way that you know the others haven't."

Sustainability has become a huge part of Apple's culture, and Tim Cook has been clear that he wants other companies and the public to know about it.

Cook is trying to set a public example not only for other companies but also for Apple's customers, a strategy that differs from Apple's typical atmosphere of secrecy. Lisa Jackson said, "We're very protective of our labs and our products, but when it comes to environmental work, Tim has been really clear that there are some secrets . . . we don't want to keep." Cook believes this transparency will inspire others to make environmentally conscious decisions in their own lives.

Apple employees are likewise proud of what their company has achieved so far. Jackson says that Cook encourages a competitive atmosphere not only surrounding Apple products but also environmental initiatives. "He doesn't let us just stop at what sounds good. We have to do what's right. And he also doesn't let us sort of buy our way out of a problem. We have to come up with a sustainable solution." Cook's determination has clearly paid off as Apple has now emerged as the one of the greenest tech companies in the world.

Chapter 9

Cook Fights the Law, and Wins

Privacy is another of Cook's values that has remained high on Apple's agenda since he took over as CEO. From the earliest mention of privacy issues in 2013 to the San Bernardino dilemma to the present day, he has taken the issue of user privacy very seriously.

Protecting the privacy of Apple users has always been a key focus for Cook, who has stated he is a "very private person" who likes "being anonymous." Under Cook, Apple has significantly bolstered the privacy controls available to users on their devices. Almost every software update released during his tenure has increased privacy protections and made it easier for consumers to ensure that their most sensitive data doesn't end up in the wrong hands, including advertisers. The expansion of privacy controls started in 2012 with iOS 6, the first iPhone and iPad update developed almost entirely under Cook's leadership.

iOS 6 introduced a dedicated Privacy menu inside the Settings app, giving users fairly simple controls over which content and data their apps had access to. The menu initially offered six sections, each containing toggles that made it easier than ever to manage permissions for each app. One section, Location Services, allowed users to block certain system services, including cellular network search, Genius for apps, and iAds, from tracking their location. iOS 6 also gave users the ability to limit ad tracking for the first time, making it more difficult for developers to deliver targeted ads based on their interests and browsing activities. This made the iPhone and iPad some of the first mobile devices to offer this protection, which has since made its way to other major platforms.

Privacy and security improvements also played an important role in iOS 7, unveiled at Apple's annual Worldwide Developers Conference in June 2013. The update's biggest talking point was its dramatic and controversial redesign, devised by Jony Ive, who had been tasked with overseeing software design following the departure of Scott Forstall in October 2012. The privacy and security improvements played second fiddle, but were just as important. Apple added support for Touch ID, its new fingerprint recognition system that debuted with the iPhone 5S. Touch ID was hailed at the time as a major step forward in security. It made securing a phone easy and negated the need to type in a passcode for every unlock, which encouraged more users to start securing their iPhones.

iOS 7 also brought Activation Lock, a feature that prevents lost or stolen devices from being wiped and reactivated without the owner's iCloud password. Activation Lock makes the iPhone and iPad significantly less appealing to would-be thieves, who quickly realized that they would not be able to sell what essentially became the world's most attractive brick as soon as it was no longer in the possession of its rightful owner. Police data published in 2014 revealed that iPhone thefts in San Francisco had fallen 38 percent since Activation Lock was made available in September 2013, while thefts in London and New York City had dropped 24 percent and 19 percent, respectively.

In November 2013, Apple published its first transparency report, which detailed the requests the company had received from government agencies seeking user data. "We believe that our customers have a right to understand how their personal information is handled, and we consider it our responsibility to provide them with the best privacy protections available," the report read. "Apple has prepared this report on the requests we receive from governments seeking information about individual users or devices in the interest of transparency for our customers around the world."

The transparency report revealed statistics on requests related to customer accounts, as well as those related to specific Apple devices. Apple promised in its first report that it would continue to advocate for greater transparency about the requests it receives, and it has followed by issuing new transparency reports every six months. "Consumer privacy is a consideration from the earliest stages of design for all our products and services," Apple continued. "We work hard to deliver the most secure hardware and software in the world." Apple also noted that its business "does not depend on collecting personal data," unlike other Silicon Valley giants such as Facebook and Google. "We have no interest in amassing personal information about our customers."

Cook took another subtle stab at companies that make money from user data in an open letter to Apple customers in September 2014. "Our business model is very straightforward: we sell great products," he wrote. "We don't build a profile based on your email content or web browsing habits to sell to advertisers. We don't 'monetize' the information you store on your iPhone or in iCloud. And we don't read your email or your messages to get information to market to you."

Apple was operating its own advertising business, iAds, at this time—but it still had no interest in mining sensitive data to make those advertisements more personal. iAds "sticks to the same privacy policy that applies to every other Apple product," Cook promised. "It doesn't get data from Health and HomeKit, Maps, Siri, iMessage, your call history, or any iCloud service like Contacts or Mail, and you can always just opt out altogether."

September 2014 also saw the release of iOS 8, which offered even greater security and privacy protections, and a new privacy policy that promised Apple would not unlock iOS devices for law enforcement agents, even those that provided a warrant. To deliver that promise, Apple cleverly changed the way in which the data on those iOS devices was encrypted. It adopted a system similar to that used by the U.S. government to protect classified military secrets, which generates an encryption key by combining your iOS passcode with a string of secret numbers that are unique to your device. It prevents others, even Apple, from decrypting your data without your passcode, so it could not unlock a device or open protected backups for government agencies even if it was forced to do so by a judge.

That encryption didn't protect everything a user had installed on their device until the release of iOS 9 in September 2015, which also brought support for content blockers that give users greater control over advertisements, cookies, and data collection tools in Safari, the default web browser on iOS. Three months earlier, Cook became the first business leader to be honored by the Electronic Privacy Information Center (EPIC) for his "corporate leadership." During his speech at the EPIC Champions of Freedom Awards Dinner, Cook reiterated Apple's commitment to protecting privacy, which the company views as a "fundamental right."

"Like many of you, we at Apple reject the idea that our customers should have to make trade-offs between privacy and security," Cook began. "We can, and we must provide both in equal measure. We believe that people have a fundamental right to privacy. The American people demand it, the Constitution demands it, morality demands it." Cook took the opportunity to again shame Facebook and Google, without naming them, for their data-driven approach. "I'm speaking to you from Silicon Valley, where some of the most prominent and successful companies have built their businesses by lulling their customers into complacency about their personal information," he continued. "They're gobbling up everything they can learn about you and trying to monetize it. We think that's wrong. And it's not the kind of company Apple wants to be."

Apple "doesn't want your data," Cook reminded attendees and Apple fans. "We don't think you should ever have to trade it for a service that you think is free but actually comes at a very high cost." He explained that Apple's dedication to protecting our data is even more important in an age when all our sensitive data, including our finances and even health information, is stored on our smartphones. "We believe the customer should be in control of their own information. You might like these so-called free services, but we don't think they're worth having your email, your search history and now even your family photos data mined and sold off for God knows what advertising purpose. And we think someday, customers will see this for what it is."

He also took the time to defend encryption and explain why Apple would not provide a "backdoor" that would allow government agencies to hack into iOS devices. "Removing encryption tools from our products altogether, as some in Washington would like us to do, would only hurt law-abiding citizens who rely on us to protect their data," he stressed. "Now, we have a deep respect for law enforcement, but on this issue we disagree. So let me be crystal clear—weakening encryption, or taking it away, harms good people that are using it for the right reasons. And ultimately, I believe it has a chilling effect on our First Amendment rights and undermines our country's founding principles." He warned, "If you put a key under the mat for the cops, the burglar can find it, too. Criminals are using every technology tool at their disposal to hack into people's accounts. If they know there's a key hidden somewhere, they won't stop until they find it."

Cook admitted during a December 2015 interview with Charlie Rose that Apple would have to comply if the government issued a warrant that demanded specific information, despite its staunch stance regarding user privacy, "because we have to by law." But thanks to the protections built into its software and devices, Apple wouldn't have much to pass over. "In the case of encrypted information, we don't have it to give," he explained. He wanted to make sure that Apple users understood that, though it would have to work within the law, Apple was committed to keeping their data safe.

Privacy Concerns

But there were a few privacy scandals over the years that made Apple users nervous. One of the first public discussions of Apple's stance on privacy took place in December 2013, when a classified document leaked, revealing that the National Security Agency had an active program, dubbed "DROP-OUTJEEP," which allowed it to eavesdrop on nearly every communication sent to or from an iPhone, using a software implant. The public was outraged, and Apple was accused of working with the NSA, allowing "backdoor" access to iOS in support of government snooping. But the company published a clear statement in response: "Apple has never worked with the NSA to create a backdoor in any of our products, including iPhone. Additionally, we have been unaware of this alleged NSA program targeting our products." The statement explained, "This functionality includes the ability to push/pull files from the device, SMS retrieval, contact list retrieval, voicemail, geolocation, hot mic, camera capture, cell tower location, etc."

The NSA claimed a 100 percent success rate on Apple devices, but Apple promised it would "continue to use our resources to stay ahead of malicious hackers and defend our customers from security attacks, regardless of who's behind them."

An earlier white paper from security researchers at QuarksLab, published in October 2013, had claimed that Apple had the ability to access iMessage conversations if it wanted to, or if it was "required to do so by a government order." But Apple had been quick to deny this too, in a statement that explained, "iMessage is not architected to allow Apple to read messages." Apple said the research "discussed theoretical vulnerabilities that would require Apple to re-engineer the iMessage system to exploit it, and Apple has no plans or intentions to do so." The company followed this with a report that outlined the government requests it had received for customers' personal data. It read:

> We believe that our customers have a right to understand how their personal information is handled, and we consider it our responsibility to provide them with the best privacy protections available. Apple has prepared this report on the requests we receive from governments seeking information about individual users of devices in the interest of transparency for our customers around the world. . . . We have reported all the information we are legally allowed to share, and Apple will continue to advocate for greater transparency about the requests we receive.

But privacy concerns continued. At a *Wall Street Journal* technology conference in Laguna Beach, California, in October 2015, Cook addressed queries about a backdoor into Apple's software, saying that "no backdoor is a must." If the NSA and other authorities have backdoor access to iOS, attackers could gain access to it, too. There is a significant risk that the backdoor would be uncovered and exploited by malicious actors, leaving hundreds of millions of iOS users at risk. After all, the ability to identify vulnerabilities in Apple's software is what spawned a once-thriving jail-breaking community that enabled unauthorized apps to be installed on iOS devices. "If someone can get into data, it is subject to great abuse," Cook added. "Strong encryption is in our nation's best interest."

Apple was making moves in the industry, but in September 2014 Cook and Apple were accused of not taking iCloud security seriously, after private photos and videos of more than a hundred celebrities were leaked online. Dubbed "Celebgate," the leak affected high-profile stars like Jennifer Lawrence, Rihanna, and Cara Delevingne—but Apple insisted its iCloud systems had not been breached. "None of the cases we have investigated has resulted from any breach in any of Apple's systems including iCloud," read a statement. "We are continuing to work with law enforcement to help identify the criminals involved." Apple said that the accounts were compromised by "a very targeted attack on user names, passwords and security questions."

iCloud "wasn't hacked," Cook reiterated during an interview with Charlie Rose. "There's a misunderstanding about this. If you think about what iCloud hacking would mean, it means somebody would get into the cloud and could fish around in people's accounts." Cook insisted, "that didn't happen." Instead, the attacker used what he called a "phishing expedition." Rather than breaking into iCloud servers, Ryan Collins of Lancaster, Pennsylvania, sent phishing emails disguised as official Apple communications that fooled users into handing over their login details. He was then able to log in to their accounts and access iPhone and iPad backups that could be used to restore the device. The hundreds of images and videos he obtained were then shared and published on the online image board 4Chan before quickly being circulated around the web. "There are lots of bad people that do this," Cook told Rose.

Although Apple didn't consider this an iCloud problem initially, backlash in the media and from fans prompted Cook to admit that the company could have done more, and he promised changes to prevent a similar fiasco. "When I step back from this terrible scenario that happened and say what more could we have done, I think about the awareness piece," he told the *Wall Street Journal.* "I think we have a responsibility to ratchet that up. That's not really an engineering thing. We want to do everything we can do to protect our customers, because we are as outraged if not more so than they are."

Cook promised Apple would begin issuing email and push notifications whenever someone tries to change an account password, restore iCloud data to a new device, or log in to an iCloud account for the first time on a new device. "We had to do things where we notify the customer quickly if it does happen," he added in his interview with Charlie Rose. "We don't want it to happen at all, but if it does, you probably want to know instantly." Prior to the leak, Apple emailed iCloud users when an account password was changed, but there were no notifications when data was restored. Apple began implementing the system just a week later, and it has worked well ever since.

Cook had been navigating these privacy issues for many years, mostly behind the scenes. But Apple's stance on privacy would come to the forefront of a national debate in 2016 after a terrorist shooting occurred in San Bernardino. The standoff would be the biggest test of Cook's career, one that almost risked the future of the company.

San Bernardino

In 2016, Tim Cook fought the law—and won.

Late in the afternoon of Tuesday, February 16, 2016, Cook and several lieutenants gathered in the "junior boardroom" on the executive floor at One Infinite Loop, Apple's old headquarters. The company had just received a writ from a U.S. magistrate ordering it to make specialized software that would allow the FBI to unlock an iPhone used by Syed Farook, a suspect in the San Bernardino shooting in December 2015 that left fourteen people dead.

The iPhone was locked with a four-digit passcode that the FBI had been unable to crack. The FBI wanted Apple to create a special version of iOS that would accept an unlimited combination of passwords electronically, until the right one was found. The new iOS could be side-loaded onto the iPhone, leaving the data intact.

But Apple had refused. Cook and his team were convinced that a new unlocked version of iOS would be very, very dangerous. It could be misused, leaked, or stolen, and once in the wild, it could never be retrieved. It could potentially undermine the security of hundreds of millions of Apple users.

In the boardroom, Cook and his team went through the writ line by line. They needed to decide what Apple's legal position was going to be and figure out how long they had to respond. It was a stressful, high-stakes meeting. Apple was given no warning about the writ, even though Cook, Apple's top lawyer, Bruce Sewell, and others had been actively speaking about the case to law enforcement for weeks.

The writ "was not a simple request for assistance in a criminal case," explained Sewell. "It was a forty-two-page pleading by the government that started out with this litany of the *horrible* things that had been done in San

Bernardino. And then this . . . somewhat biased litany of all the times that Apple had said no to what were portrayed as very reasonable requests. So this was what, in the law, we call a speaking complaint. It was meant to from day one tell a story . . . that would get the public against Apple."

The team came to the conclusion that the judge's order was a PR move—a very public arm twisting to pressure Apple into complying with the FBI's demands—and that it could be serious trouble for the company. Apple "is a famous, incredibly powerful consumer brand and we are going to be standing up against the FBI and saying in effect, 'No, we're not going to give you the thing that you're looking for to try to deal with this terrorist threat,'" said Sewell.

They knew that they had to respond immediately. The writ would dominate the next day's news, and Apple had to have a response. "Tim knew that this was a massive decision on his part," Sewell said. It was a big moment, "a bet-the-company kind of decision." Cook and the team stayed up all night—a straight sixteen hours—working on their response. Cook already knew his position—Apple would refuse—but he wanted to know all the angles: What was Apple's legal position? What was its legal obligation? Was this the right response? How should it sound? How should it read? What was the right tone?

Cook was very concerned about the public's reaction and knew that one of the outcomes of his action could be that Apple would be accused of siding with terrorists. What kind of company wouldn't help the FBI in a terrorist investigation? From a public relations standpoint, Apple had always been on the side of privacy advocates and civil libertarians. This case put the company unexpectedly on the side of a terrorist. This was brand-new territory, and Cook had to figure out how to navigate it. He had to show the world that he was advocating for user privacy rather than supporting terrorism.

At 4:30 a.m., just in time for the morning news cycle on the East Coast, Cook published an open letter to Apple customers explaining why the company would be opposing the ruling, which "threatens the security of our customers." He referenced the danger that could come from the government having too much power: "The implications of the government's demands are chilling," he wrote. "If the government can use the All Writs Act to make it easier to unlock your iPhone, it would have the power to reach into anyone's device to capture their data."

Apple had been working with the FBI to try to unlock the phone, providing data and making engineers available, Cook explained. "But now the U.S. government has asked us for something we simply do not have, and something we consider too dangerous to create . . . a backdoor to the iPhone." He continued, "In the wrong hands, this software—which does not exist today—would have the potential to unlock any iPhone in someone's physical possession." This could have potentially disastrous consequences, leaving users powerless to stop any unwanted invasion of privacy. "The FBI may use different words to describe this tool, but make no mistake: Building a version of iOS that bypasses security in this way would undeniably create a backdoor. And while the government may argue that its use would be limited to this case, there is no way to guarantee such control."

Cook then accused the government of trying to force Apple "to hack our own users and undermine decades of security advancements that protect our customers . . . from sophisticated hackers and cybercriminals." It would be a slippery slope from there. The government could then demand that Apple build surveillance software to intercept messages, access health records or financial data, or track users' locations. Cook needed to draw a line. He believed the FBI's intentions were good, but it was his responsibility to protect Apple users. "We can find no precedent for an American company being forced to expose its customers to a greater risk of attack," he wrote. Though it was difficult for him to resist orders from the U.S. government, and he knew he'd face backlash, he needed to take a stand.

Long-Running Debate

The magistrate's order thrust into the spotlight a long-running debate Apple had been having with the authorities about encryption. Apple and the government had been at odds for more than a year, since the debut of Apple's encrypted operating system, iOS 8, in late 2014.

iOS 8 added much stronger encryption than had been seen before in smartphones. It encrypted all the user's data—phone call records, messages, photos, contacts, and so on—with the user's passcode. The encryption was so strong, not even Apple could break it. Security on earlier devices was much weaker, and there were various ways to break into them, but Apple could no longer access locked devices running iOS 8, even if law enforcement had a valid warrant. "Unlike our competitors, Apple cannot bypass your passcode and therefore cannot access this data," the company wrote on its website. "So it's not technically feasible for us to respond to government warrants for the extraction of this data from devices in their possession running iOS 8."

The update had repeatedly stymied investigators. At the New York press event two days after Cook's letter on San Bernardino, the authorities said that they had been locked out of 175 iPhones in cases they were pursuing. For more than a year, law enforcement at the highest levels had been pressuring Apple for a solution. "When the FBI filed in San Bernardino, I think many people in the public perceived that as the beginning of something," said Sewell. "Whereas in reality, it was a long point leading up to that, with a lot of activity that preceded the actual decision by [FBI director James] Comey to file."

Sewell explained that he, Cook, and other members of Apple's legal team had been meeting regularly with heads of the FBI, the Justice Department, and the attorney general in both Washington and Cupertino. Cook, Sewell, and others had met not only with James Comey, but also with Attorney General Eric Holder, Attorney General Loretta Lynch, FBI director Bob Mueller (Comey's predecessor), and Deputy Attorney General Sally Yates.

Cook and Sewell met with Eric Holder and Jim Cole, then the deputy attorney general, in late 2014, and FBI agents told them they were "interested in getting access to phones on a mass basis." This was way before the attack in San Bernardino, and Apple made it clear from the start that they were not going to grant the FBI access to hack into Apple users' phones. Cook and Sewell told Holder and Cole that they "didn't think that that was an appropriate request to be made of a company that has as its primary concern the protection of all citizens." They had a similar conversation with Lynch and Yates.

Sewell said that during the discussions, it was clear that some law enforcement officials weren't convinced by the broader social issues. Some were intellectually sympathetic to their position, but as officers of the law, they insisted they needed access to pursue cases. But Sewell said Cook stuck to his position that security and privacy was a cornerstone. Cook was adamant that any attempt to bypass security would be very dangerous. Once a backdoor had been created, it could easily be leaked, stolen, or abused.

But when the San Bernardino case came along, law enforcement saw it as an opportunity to force Apple's hand. "There was a sense at the FBI level that this is the perfect storm," said Sewell. "We now have a tragic situation. We have a phone. We have a dead assailant. This is the time that we're going to push it. And that's when the FBI decided to file [the writ ordering Apple to create a backdoor]."

The Firestorm

As Cook and his team had predicted, the judge's order ignited a firestorm in the media. The story dominated the news all week and would continue to be headline news for two months. Apple's response drew strong condemnation from law enforcement, politicians, and pundits, like Democratic senator Dianne Feinstein of California, head of the U.S. Senate Intelligence Committee, who called on Apple to help with the "terrorist attack in my state" and threatened legislation.

At a press conference in Manhattan, William Bratton, New York City police commissioner, also criticized Apple's policy. He held up a phone involved in a separate investigation of the shooting of two police officers. "Despite having a court order, we cannot access this iPhone," he told the assembled journalists. "Two of my officers were shot, [and] impeding that case going forward is our inability to get into this device."

A few days later, Donald Trump, then a presidential candidate, called for a boycott against Apple at a campaign rally in Pawleys Island, South Carolina. Trump even accused Cook of being politically motivated: "Tim Cook is looking to do a big number, probably to show how liberal he is." Trump was playing to his conservative audience, trying to make Cook seem like a liberal bad guy and using scare tactics to make it seem like Apple was siding with terrorists. He tweeted further attacks on Apple, calling again for a boycott until the company handed over the information to the FBI.

With so many politicians and officials against Apple, the American public lined up against it, too. A Pew survey found that 51 percent of people said Apple should unlock the iPhone to help the FBI, with only 38 percent supporting Cook's position. But a few days later, another poll by Reuters/ Ipsos came to a different conclusion. According to that poll, 46 percent agreed with Apple's stance, 35 percent disagreed, and 20 percent didn't know. The difference was attributed to the phrasing of the question: The Pew survey question gave less information about Apple's position and appeared to be biased toward the FBI. An analysis of the emojis used in social media came to a similar mixed conclusion. By analyzing positive and negative emojis in people's tweets (smiley faces, frowns, claps, thumbs up, and thumbs down), a marketing firm called Convince & Convert found a fairly even split between those who sided with Apple and those who supported the FBI. Though this approach was less than scientific, it was clear the public was divided. This experience was unprecedented, and many did not know what to think.

And ultimately, it wasn't all bad. Cook's stance also appeared to have some influence on public opinion. In hundreds of responses to Trump's tweets, lots of citizens defended Apple's actions. Trump's tweets tended to bring out contrarian opinions, but most reactions tended toward defenses of Apple. One responder tweeted, "Boycotting Apple products is absurd. Break into one phone, none of us will have privacy. The govt can't be trusted!!"

Several high-profile figures also voiced support for Cook and Apple, including Facebook CEO Mark Zuckerberg, Google CEO Sundar Pichai, Twitter CEO Jack Dorsey, and Edward Snowden, the NSA whistleblower. The *New York Times* editorial board also weighed in on Apple's side. In an editorial titled "Why Apple Is Right to Challenge an Order to Help the F.B.I.," they wrote, "There's a very good chance that such a law, intended to ease the job of law enforcement, would make private citizens, businesses and the government itself far less secure." Cook and his team obviously agreed, and hunkered down to continue the fight.

The War Room

For the next two months, the executive floor at One Infinite Loop turned into a 24/7 situation room, with staffers sending out messages and responding to journalists' queries. One PR rep said that they were sometimes sending out multiple updates a day with up to seven hundred journalists cc'd on the emails. This is in stark contrast to Apple's usual PR strategy, which consists of occasional press releases and routinely ignoring reporters' calls and emails.

Cook also felt he had to rally the troops, to keep morale high at a time when the company was under attack. In an email to Apple employees, titled "Thank you for your support," he wrote, "This case is about much more than a single phone or a single investigation." He continued, "At stake is the data security of hundreds of millions of law-abiding people and setting a dangerous precedent that threatens everyone's civil liberties." It worked. Apple employees trusted their leader to make the decision that was right not only for them but also for the general public.

Cook was very concerned about how Apple would be perceived throughout this media firestorm. He wanted very much to use it as an opportunity to educate the public about personal security, privacy, and encryption. "I think a lot of reporters saw a new version, a new face of Apple," said the PR person, who asked to remain anonymous. "And it was Tim's decision to act in this fashion. Very different from what we have done in the past. We were sometimes sending out emails to reporters three times a day on keeping them updated."

Outside Apple's walls, Cook went on a charm offensive. Eight days after publishing his privacy letter, he sat down for a prime-time interview with ABC News. Sitting in his office at One Infinite Loop, he sincerely explained Apple's position. It was the "most important [interview] he's given as Apple's CEO," said the *Washington Post*. "Cook responded to questions with a raw conviction that was even more emphatic than usual," wrote the paper. "He used sharp and soaring language, calling the request the 'software equivalent of cancer' and talking about 'fundamental' civil liberties. He said he was prepared to take the fight all the way to the Supreme Court." It was clear that Apple's leader wouldn't back down from his beliefs, even when things got really tough.

The interview went well, and back at Apple's HQ, staffers in the war room felt it was a pivotal point. They thought Cook did a great job not only explaining Apple's point of view but also showing the world that he was a compassionate, ethical leader whom users could trust to maintain their privacy. "This is not a rapacious corporate executive who's out to make a bunch of money," said Sewell. "This is somebody who you could trust. Somebody who does what he says he's going to do. And doesn't do things that are malicious or that are ill-intentioned but tries to be fair, tries to be a good steward of the company and means what he says and does things that he believes in." Apple employees had known this side of Tim Cook for many years, but the public was getting a glimpse for the first time. This was a victory for Apple, since many members of the public did not initially approve of Apple's decision to keep iPhone information away from the FBI.

Apple won another victory at the end of February, when a court in New York rejected an FBI request to order Apple to open the phone of a minor drug dealer. Judge James Orenstein agreed with Apple's position that the All Writs Act could not be used to order the company to open its products. "The implications of the government's position are so far-reaching—both in terms of what it would allow today and what it implies about Congressional intent in 1789," he said.

Although this particular case wasn't binding on the court in San Bernardino, Sewell said it gave the company much-needed ammunition with the press. "For us it was very, very important," he said. "It enabled us to then go back to the press and go back to people who had generally been detractors and say, 'This isn't about Apple commercialism. This isn't about Apple being a bad actor. This is a principled position and the only judge in the country that's looked at this agreed with us.'" Cook and Sewell felt confident that with Judge Orenstein on their side, others would soon be, too.

No Privacy in America

As the battle raged on, support from privacy advocates grew, but public opinion on Apple's decision was still largely divided. An NBC survey of twelve hundred Americans conducted in March 2016 found that 47 percent of respondents believed the company should not cooperate with the FBI, while 42 percent believed it should. Forty-four percent of respondents said they feared the government would go too far and violate the privacy of its citizens if Apple were to meet its demands.

The United Nations voiced its support for Apple, with special rapporteur David Kaye arguing that encryption is "fundamental to the exercise of freedom of opinion and expression in the digital age." Kaye continued by stating that the FBI's "order implicates the security, and thus the freedom of expression, of unknown but likely vast numbers of people, those who rely on secure communications." But the FBI continued its PR offensive, with then director James Comey telling attendees at a Boston College conference on cybersecurity in March that "there is no place outside of judicial reach. . . . There is no such thing as absolute privacy in America."

The lowest point for Apple was when Attorney General Loretta Lynch criticized the company during a keynote speech at the security-oriented

RSA Conference in San Francisco. Lynch essentially accused Apple of defying the law and the courts. Her comments were widely reported and featured on the evening news. "Nothing could be further from the truth," Sewell said. "For the attorney general to go on public television and say, 'Apple is in breach of a court order and is therefore acting unlawfully,' is inflammatory. . . . A lot of media picked up this as the attorney general saying that Apple is . . . disregarding a court order. But there was no court order." The judge's writ requested Apple's help in the case; it did not compel the company to do so, a distinction that was lost—or ignored—by many critics. Apple wasn't breaking any laws, and it was determined to fight for user privacy, despite lots of pressure from the government.

The Case Is Dropped

Six weeks after the judge filed the motion against Apple, on March 28, Sewell and the legal team flew down to San Bernardino to argue their case before the judge. Cook was preparing to fly down the next day to testify.

But that evening, the FBI backed down, asking the court to indefinitely suspend the proceedings against Apple. The FBI said it had successfully accessed the data stored on the phone, though it didn't explain how. It was later revealed that the FBI had gained access to Farook's iPhone with the help of Israeli phone forensics company Celebrite. At a Senate Judiciary hearing in May, Senator Dianne Feinstein revealed that it had cost the FBI $900,000. Officials had previously admitted that the FBI didn't find any information they didn't already have, and no evidence of contacts with ISIS or other supporters. The FBI had to drop the fight with Apple, Sewell explained, because its entire position was that it couldn't access the iPhone without Apple's help. When it turned out that they could in fact access the phone, the case collapsed.

Privacy advocates celebrated the end of the case and Apple's apparent victory. "The FBI's credibility just hit a new low," said Evan Greer, campaign director for Fight for the Future, an activist group that promotes online privacy. "They repeatedly lied to the court and the public in pursuit of a dangerous precedent that would have made all of us less safe. Fortunately, internet users mobilized quickly and powerfully to educate the public about the dangers of backdoors, and together we forced the government to back down."

But Cook was personally disappointed that the case didn't come to trial. Even though Apple had "won" and wouldn't be forced to create the backdoor, nothing had really been resolved. "Tim was a little disappointed that we didn't get a resolution," said Sewell. He "really felt it would have been fair and it would have been appropriate for us to have tested these theories in court. . . . [Though] the situation that was left at the end of that was not a bad one for us, he would have preferred to go ahead and try the case." The issue remains unresolved to this day. It could be reawakened at any time, and under the Trump administration it is probably likely to be. It was just another skirmish in the war for privacy and security, and as technology evolves, the battle is likely to erupt again in the future.

Cook Doubles Down on Privacy

Apple continued to improve on its privacy protections with the launch of iOS 11.3 in April 2018, which added a new icon that makes it explicitly clear to users when their personal data is being collected by an Apple service. "You won't see this with every feature since Apple collects this information only when needed to enable features, secure our services, or personalize your experience," explains a notice that greets users after updating to iOS 11.3. "Apple believes privacy is a fundamental human right, so every Apple product is designed to minimize the collection and use of your data, use on-device processing whenever possible, and provide transparency and control over your information."

The release of iOS 11.3 coincided with growing controversy surrounding Facebook's privacy practices after it was revealed that Cambridge Analytica, a British political consulting firm, had been collecting Facebook user data to produce strategic communications in an effort to influence voter opinion during the Trump presidential campaign. Facebook confirmed that up to eighty-seven million of its users were affected by Cambridge Analytica's "inappropriate" data collection practices, and it apologized for not doing more to monitor third-party developers. The scandal sparked something of a spat between Cook and Facebook CEO Mark Zuckerberg, who clearly have very different opinions on how user data should be handled.

"I think that this certain situation is so dire and has become so large that probably some well-crafted regulation is necessary," Cook said during an appearance at the China Development Forum in Beijing in late March. "The ability of anyone to know what you've been browsing about for years, who your contacts are, who their contacts are, things you like and dislike and every intimate detail of your life—from my own point of view, it shouldn't exist." When Recode's Kara Swisher asked Cook what he would do in Zuckerberg's situation as Congress demanded answers to its concerns, he simply replied, "I wouldn't be in this situation." And so far, that has remained true. Apple users can rest assured that Apple is not using their data in the same way that Facebook has been, because Apple has a leader who strongly values user privacy.

Chapter 10

Doubling Down on Diversity

After three years at the helm as Apple's CEO, Cook was coming into his own as the company's leader. He was more relaxed in public, speaking off-the-cuff and cracking jokes in interviews, and happily posing for selfies everywhere he went. His colleague Greg Joswiak noted that everyone was starstruck—even famous athletes and film stars asked for pictures with him. "Even among celebrities Tim seems to be the big celebrity, but he doesn't act like it," Joswiak said. "It just never goes to his head. He just never acts like a celebrity. He acts like Tim."

Away from the public glare, Cook's six values were changing Apple's culture. The company had never been more devoted to promoting diversity, equality, education, and accessibility. The year 2014 was a momentous one for Cook as CEO, but it would go down in history for a very special reason.

On October 30, Cook wrote a heartfelt essay for Bloomberg, titled "Tim Cook Speaks Up," in which he revealed publicly for the first time that he was gay. There had been rumors about his sexuality before, and Cook says that he had been open with "plenty" of colleagues at Apple—but this was the first time he confirmed it to the world at large. "While I have never denied my sexuality, I haven't publicly acknowledged it either, until now," he wrote. "So let me be clear: I'm proud to be gay, and I consider being gay among the greatest gifts God has given me." Cook, now the first CEO of a Fortune 500 company to come out, continued that "being gay has given me a deeper understanding of what it means to be in the minority and provided a window into the challenges that people in other minority groups deal with every day."

Being gay, Cook explained, made him more empathetic and gave him the confidence to be himself, follow his own path, and rise above adversity and bigotry. "It's also given me the skin of a rhinoceros, which comes in handy when you're the CEO of Apple." He noted how the world has changed so much since he was a child, with America moving toward marriage equality, and with other public figures coming out to make the country's culture "more tolerant." He said he didn't consider himself an activist but realized how much he had benefited from the sacrifice of others. "So if hearing that the CEO of Apple is gay can help someone struggling to come to terms with who he or she is, or bring comfort to anyone who feels alone, or inspire people to insist on their equality," he said, "then it's worth the trade-off with my own privacy."

Cook later told *The Late Show* on CBS that he decided to make his sexuality public after he realized he could help America's queer youth. "Kids were being bullied in school, kids were getting discriminated against, kids were even being disowned by their own parents. I needed to do something," he said. "I valued my privacy significantly, [but] I felt that I was valuing it too far above what I could do for other people. I wanted to tell everyone my truth."

Cook's decision to disclose his sexuality, four months after leading eight thousand Apple employees at the forty-third annual San Francisco Pride Parade, was big news among Apple fans and the wider tech community, with the reaction almost entirely positive. "From one son of the South and sports fanatic to another, my hat's off to you," tweeted former president Bill Clinton. "Inspirational words from Apple CEO Tim Cook on being gay, and standing up for equality," added billionaire entrepreneur Richard Branson. Said Lloyd Blankfein, chairman and CEO of Goldman Sachs, "He's chief executive of the Fortune One. Something has consequences because of who does it, and this is Tim Cook and Apple. This will resonate powerfully." Bob Iger, CEO of the Walt Disney Company, said, "Tim sacrificed his own privacy to ensure a generation of young people understand that they matter, whoever they are." The *Huffington Post* declared that Cook had "just changed America in a way Steve Jobs never could." Since there were no other publicly out Fortune 500 CEOs, one could have expected that Cook's revelation might have been taken badly. But in reality, there wasn't much of a reaction at all. Cook's essay was never considered a major bombshell.

Lisa Jackson, Apple's vice president of environment, policy, and social initiatives, got an email from Cook the night before his op-ed was published. He wanted her to see it first before it hit newsstands. "I felt like I was witnessing an act of service and bravery and courage and leadership that was as profound as any I had witnessed in my life," she said, clearly moved. "You know what that meant to so many people. What that meant to so many kids. You know young people, teenagers, students, people who didn't feel like they had somebody looking out for them. . . . Tim's a fighter and having him on your side is definitely . . . a good thing."

For every tweet, Facebook status, or column celebrating the news, there were twice as many lighthearted jokes. "Would have been more interesting if Tim Cook revealed that he uses Windows," said one Twitter user. And Samsung quips ruled the roost. "Samsung just announced that their next CEO will be gay!" joked David Wolf, adding, " 'Our gay CEO will be 25% gayer than Tim Cook,' said Samsung PR."

Wall Street wasn't fazed by Cook's essay either, with Apple stock in virtually exactly the same condition in premarket trading immediately after his news broke, dipping less than 1 percent after the market officially opened. "To put it another way: The CEO of the world's most valuable company publicly announced he's a gay man, and Wall Street could not care less," wrote Seth Fiegerman for Mashable. Apple analyst Gene Munster described the news as "a non-event because Tim has already proven he's an excellent CEO." In an opinion piece for CNBC, coauthored by fellow Piper Jaffray analyst Doug Clinton, Munster continued, "It may seem strange to have Wall Street analysts weigh in on such a personal topic and we recognize that the announcement is bigger than shares of AAPL. But Cook's decision to announce that he is gay had the potential to impact the stock and many wondered if it would. We were happy to see investors vote through a roughly unchanged stock price that Tim Cook is just as capable a CEO today as he was before the announcement."

It is not known if Cook is in a relationship. He's never been photographed with a partner by paparazzi, and there are few rumors about his private life. One rumor—that he's in a relationship with a Silicon Valley venture capitalist—was easily debunked from the VC's Instagram feed (which is now private). There were numerous photos of the man with his partner, and it wasn't Tim Cook. Cook appears to live alone in Palo Alto. If he is in a relationship, it must be very difficult for him to conduct it privately. One wonders if he eats out at restaurants, or goes to parties or on vacation. While researching this book, I didn't pry into his personal life at all. Cook keeps his private life private, and I'm happy to respect that.

Almost a year after his Bloomberg essay, Cook was awarded a Visibility Award at the nineteenth annual national dinner hosted by the Human Rights Campaign, America's leading LGBTQ civil rights organization. "Tim Cook is a visionary whose leadership of Apple has been nothing short of remarkable," said HRC president Chad Griffin. "His willingness to bravely and directly speak his truth has not only given hope to countless people around the world, it has saved lives. Through his example and Apple's commitment to equality, LGBTQ young people in particular can look to Tim Cook's incredible career and know that there is nothing holding them back. They can dream as big as their minds allow them to, even if they want to be the CEO of one of the world's largest companies." The HRC commended Cook for his impact on "Corporate America and beyond," for making Apple one of the earliest major supporters of the Equality Act, and for lending his voice to fight for full LGBTQ equality.

"I am proud to be a part of this community," Cook told a welcoming crowd at the HRC's national dinner. "As some of you might remember, I wrote an essay that was deeply personal. I wanted to lend my voice to people who might not be ready to exercise theirs. It was an open letter to the public, but it was addressed most of all to everyone who had been rejected by their friends, or communities, or their families, simply because of who they are." Cook stated that he did not write the essay for attention, though there was never a suggestion that he might have, and added that despite being a private person by nature, "sometimes you just have to be loud. Because people need to hear that being gay is not a limitation," he said, beaming. "People need to hear that being gay doesn't restrict your options in life. People need to hear that you can be gay or transgender and be whatever else you want to in life."

During his acceptance speech, Cook revealed that the response to his essay over the past year had been overwhelming, with people reaching out from all over the world to talk about their struggle, and others wishing they could do more to help their loved ones. Some said that he inspired them to share their sexual orientation for the first time. "Some of the most touching notes I've received are from parents who love their children more than anything and can't bear to see them struggle for acceptance," he continued. "Some of the most hopeful notes are from folks who are just happy to see the world changing for the better." He shared one of the messages he had received, from a Vietnam veteran in Oregon, with the audience. "Tim, I hope that someday people will look at announcements such as yours and greet them with a yawn," it read. "We should simply accept people for who they are as citizens, for being good human beings, and for enhancing our lives and doing their best. Isn't that the American way?"

Cook agreed, but he acknowledged that America has a long way to go before all its citizens can enjoy equal protection. Despite the changes in American society, with the Supreme Court legalization of same-sex marriage, and TV shows like *Will & Grace* and *Modern Family* bringing gay characters into the mainstream, the United States still had a lot of work to do. Cook went on to shame the thirty-one U.S. states that at the time still had no laws to protect gay and transgender people from discrimination and no state legal protection from being fired or being evicted for "who you are and who you love." He blasted parents who send kids away for reparative therapy in pursuit of "a cure" and bullies who push others to the brink of suicide. He also acknowledged that the LGBTQ community wasn't the only one facing inequalities. "The way I look at this is simple—discrimination against anyone holds everyone back," he said, to cheers and applause from the crowd. "And as all of us know, discrimination doesn't simply fade. It doesn't recede of its own accord. It has to be pushed back, challenged, overcome, and then kept at bay. That requires determination. That requires vigilance." He concluded that advocates and activists like

HRC, and global companies like Apple, all have a role to play in the fight for equality. "For together, we will pave the sunlit path to justice."

Former vice president Joe Biden, who gave the keynote address at the HRC national dinner, accompanied by a record number of LGBTQ appointees from the Obama administration, praised Cook's decision to make his sexuality public. Biden described Cook as a man "who's turned the world upside down, who understands that equality is not only a moral imperative, but is the heart of our economic might and our dynamism." Cook "has given so much encouragement to so many brilliant young women and men in the LGBTQ community," he added.

Greg Joswiak, Apple's vice president of worldwide product marketing, said Cook's coming out was the turning point in his stewardship of Apple. Internally, Apple staff had seen he was an exceptional leader, but the world hadn't yet woken up to it. Cook's coming out was the act of leadership that clearly told the world he was his own man. "The world was starting to respect the job he was doing," Joswiak added. "And to me that was a turning point in there. . . . I think that's when the world recognized it."

Person of the Year

In light of his and Apple's achievements in 2014, Cook was named Person of the Year by the *Financial Times*. "Financial success and dazzling new technology alone might have been enough to earn Apple's steely chief executive the *FT*'s vote as the 2014 Person of the Year, but Mr. Cook's brave exposition of his values also sets him apart," explained the newspaper's Tim Bradshaw and Richard Waters.

The *Financial Times* praised Cook for holding his nerve through attacks from activist investors and a loss of faith among some who were adamant that Apple could not succeed without Jobs; for championing diversity, sustainability, and supply chain transparency—but most of all for publicly revealing his sexuality for the first time. "It was a rare glimpse into his closely guarded personal life that also put at risk Apple's brand in less tolerant parts of the world," the paper said, referring to Cook's personal essay. "His eloquent defense of equality came after a year of faltering progress on gay marriage in the U.S. and as arguments rage about the lack of diversity among the people running the Silicon Valley companies, including Apple, who shape so much of our culture."

Cook was also credited for having added three women to what was once an all-white-male Apple executive team, and for changing Apple's board charter to add a commitment to seeking out candidates from minorities when appointing directors. "I thought it would be impossible to replace Steve, and to some extent, that's true," professor Michael Cusumano of MIT's Sloan School of Management admitted to the *Financial Times*. "But internally the spirit is still alive and the company is organizing around a less confrontational culture. We have to give Tim credit for that."

Equality and Diversity Are Good for Business

Cook's sexuality obviously influenced his outlook on equality and diversity. It often takes an outsider to be a champion for other outsiders. Cook was creating a less confrontational culture, and a fairer one where equality and diversity were championed. In November 2013, he penned his first op-ed as Apple CEO for the *Wall Street Journal*. Titled "Workplace Equality Is Good for Business," it reiterated Cook's commitment to "creating a safe and welcoming place for all employees, regardless of their race, gender, nationality or sexual orientation."

"Long before I started work as the CEO of Apple, I became aware of a fundamental truth," Cook began. "People are much more willing to give of themselves when they feel that their selves are being fully recognized and embraced." He urged senators to support the Employment Non-Discrimination Act to prohibit employment discrimination based on actual or perceived sexual orientation or gender identity. "So long as the law remains silent on the workplace rights of gay and lesbian Americans, we as a nation are effectively consenting to discrimination against them," he concluded. "Congress should seize the opportunity to strike a blow against such intolerance by approving the Employment Nondiscrimination Act." The bill passed the Senate in late November 2013, with bipartisan support, by a vote of 64–32.

Innovation Through Diversity

Workplace diversity—one of Cook's core values—is also part of his innovation strategy. He believes that a diverse workforce is not just a good in its own right, but also something that will help Apple to innovate, by bringing a variety of voices and experiences to the product development process. In fact, he's put it in even stronger terms: Diversity is "the future of our company," he said in 2015. Cook has said that Apple is a "better company" and creates the best products by being more diverse in its experience, knowledge, and viewpoints. "One of the reasons Apple products work really great . . . is that the people working on them are not only engineers and computer scientists, but also artists and musicians. It's this intersection of the liberal arts and humanities with technology that makes products that are magical."

Companies like Apple that straddle the globe need people who can work with coworkers from all over the world, and also serve customers from all over the world, Cook explained in an interview with the *Auburn*

Plainsman, the student newspaper of his alma mater. "The world is intertwined today, much more than it was when I was coming out of school," he said. "Because of that, you really need to have a deep understanding of cultures around the world." He added, "I have learned to not just appreciate this but celebrate it. The thing that makes the world interesting is our differences, not our similarities."

Cook has placed value in a diverse executive team, describing to Charlie Rose how his exceptionally talented team, who are "capable of doing incredible things," were helping him and the company achieve great success because of their differences. He pointed to Jony Ive, Craig Federighi, Jeff Williams, Dan Riccio, and newly appointed retail chief Angela Ahrendts, and said, "It's a privilege of a lifetime to work with them." He noted how these executives, each with different talents, were complementing his own. "I believe in diversity with a capital D," he said to Rose. "And that's diversity in thought and diversity in any way you want to measure it. And so the people that surround me are not like me. They have skills that I don't have." He continued, "Everybody is a functional expert. And then we, collectively, to get things done, work together as a team. Because the work really happens horizontally in our company, not vertically." He admitted that Apple executives argue and debate, and don't always agree on everything. "But we have great respect for one another and we trust one another and we complement one another. And that makes it all work."

Though his executive team may be diverse in thought, it lacks true diversity in race and gender. Cook's talk of "diversity with a capital D" is a dodge to avoid the obvious fact that Apple's leadership is still overwhelmingly white and male. Cook has made strides in adding more women and people of color to Apple's upper management, but it's clearly not enough. He seems sincere in his efforts to change Apple and the wider tech industry—

and it's true that these changes take time—but Apple could and should be doing far more to become more diverse.

Cook has been actively encouraging the employment of underrepresented minorities, like the disabled and veterans, and he believes that not enough leaders are speaking out about diversity. He quoted Dr. King's "the appalling silence of the good people," saying that part of the problem is that people with good intentions don't speak up. It's not a subject a lot of CEOs engage in. Speaking up can be difficult "because society unfortunately rewards the keep your head down approach . . . but doing that won't move . . . the country forward, industries forward or companies forward. You don't solve diversity like that." The solution involves speaking up about the problem, as well as instituting programs to tackle it. "I try to look at myself in the mirror and ask myself if I'm doing enough," he said. "And if the answer is no, I try to do something more."

Cook is optimistic that change will occur. "I'm convinced we're going to move the dial," he told Mashable. "It's not an overnight thing, we . . . know that. But it's also not an unsolvable issue. It's readily solvable. Because most of the issues have been created by humans, so they can be fixed." And he's doing everything in his power to fix it. "If you believe as we believe that diversity leads to better products, and we're all about making products that enrich people's lives, then you obviously put a ton of energy behind diversity the same way you would put a ton of energy behind anything else that is truly important."

Promoting Women

Cook has often expressed concern about the number of women in the tech field. "I think the US will lose its leadership in technology if this doesn't change," he told the *Plainsman*. "Women are such an important part of the workforce. If STEM-related fields continue to have this low representation of women, then there just will not be enough innovation in the United States. That's just the simple fact of it."

Cook said it's a "cop-out" when the tech industry shrugs its collective shoulders and says women aren't interested in careers in tech. Instead, the onus, he said, is on the industry to attract more women. "I think it's our fault . . . the whole tech community," he said. "I think in general we haven't done enough to reach out and show young women that it's cool to do it and how much fun it can be."

One important way to encourage greater participation in STEM is to have more female role models. Until recently, Bozoma Saint John was one of Apple's highest-profile female leaders. At Apple, she had a very public profile. Seeing women in leadership positions in tech is important, Cook has said. Representation matters.

During his tenure as CEO, Cook has been proactive about increasing diversity at Apple. He has promoted and recruited women and minorities to Apple's executive ranks. In September 2011, he promoted Cuban American Eddy Cue to senior vice president of Internet software and services. He hired Lisa Jackson, the first African American to head up the EPA, to run environmental efforts in May 2013. In October of that same year, he hired Angela Ahrendts, the former CEO of Burberry, to oversee Apple's retail stores. In 2014, he promoted Denise Young Smith as the head of human resources, and in May 2017 promoted her again to VP of diversity and inclusion. That same year, the founding partner and director of Black-Rock, Sue Wagner, was appointed to Apple's board of directors.

Cook also instituted an annual "inclusion and diversity" report akin to the company's annual environmental and supplier responsibility reports. These high-profile, public reports signal how serious Apple is about these initiatives under Cook. They seem designed to keep the company and executives' toes to the fire.

During Cook's tenure, Apple has been increasing the number of people of color in its advertising and marketing campaigns. All of Apple's ads and marketing materials feature a diverse cast. Apple has also increased the number of women appearing in its keynotes. Whereas in the Jobs era it was mostly Steve Jobs onstage, under Cook there are lots more Apple staffers, frequently women, brought out to introduce new products. A snarky post on Quartz in 2015 showed a chart of all the times women had been onstage during Apple keynotes in the previous two years—0, 0, 0, 0, 0, 0, and 0. But after Cook said he "totally" agreed that more women needed to be onstage, his WWDC keynote in June 2018 featured six women presenters, all of whom worked in leadership positions at Apple.

Cook appears to be very serious about boosting diversity, but unfortunately, the pace of change is glacial because it depends on not only changing people's attitudes, but also on workforce turnover. Most of Apple's current leaders have been in place for decades and will likely remain so for years to come. The same goes for rank-and-file staffers. Many of Apple's staff will be in their jobs for twenty years or more, and it might take decades for the workforce to change its makeup. Change is happening, but it's not fast enough.

Apple's Makeup

Apple's first diversity report was released in August 2014, revealing what most people already knew: Apple is largely white and male. In 2014, 70 percent of Apple's global workforce were men, and only 30 percent were women. In the United States, 55 percent of Apple staffers were white, with Asians making up the second largest ethnicity of the workforce at

15 percent. Hispanics accounted for 11 percent, and 7 percent identified as black (compared to around 13 percent in the general population). Those who did not declare an ethnicity accounted for the remaining 12 percent. Unfortunately, most of the staffers from underrepresented minorities worked in Apple's retail stores, not in well-paid engineering jobs or in management or the executive suite.

Cook expressed disappointment with the results. "As CEO, I'm not satisfied with the numbers on this page," he said in a statement on the diversity report website. "They're not new to us, and we've been working hard for quite some time to improve them. We are making progress, and we're committed to being as innovative in advancing diversity as we are in developing our products." He reaffirmed his commitment to increasing diversity at Apple.

But things haven't improved much in the last few years. In November 2017, women had made only slight gains, while minority groups lost ground. In the United States, Apple's tech workforce—the rank and file, including its retail workers—were 52 percent white and 77 percent male. The year before it was at 55 percent white and 77 percent male. The proportion of underrepresented minorities—blacks, Hispanics, and multiracial people—declined slightly from 18 percent to 17 percent.

Meanwhile, Apple's leadership remains predominantly white and male. According to the report, Apple's leadership is 71 percent male, and 66 percent white (down just 1 percentage point from the previous year); 23 percent Asian (up from 21 percent the previous year). The percentage of blacks, Hispanics, and multiracial people didn't change. Apple's leadership page illustrates the lack of diversity at the very top. Of seventeen executives shown, eleven are white males and three are white females. There's one African American woman, one Hispanic man, and one Asian woman. But Apple said it was trying to accelerate change, and half of the people hired in the year covered by the report (July 2016 to July 2017) were either women or members of minority groups.

Sadly, Apple actually beats some of its rivals in Silicon Valley. Most of Apple's peers—Facebook, Intel, Google, Twitter, and Microsoft—are also all about 50 percent white, 30–40 percent Asian, and the rest similar percentages of Hispanic and black workers (about 10 percent). While Apple has a larger percentage of Hispanic and black employees, most of those jobs are in lower-paying retail, jobs that are also often part-time, with fewer benefits and opportunities to move up into better jobs.

Shareholder Pressure

Apple has been under pressure from shareholders to increase diversity, particularly at the executive level. In May 2017, Cook promoted Apple veteran Denise Young Smith to VP of diversity and inclusion, but she was only on the job for less than a year. After a long and successful career at Apple, she left under mysterious circumstances, after making a strange and surprising comment at a conference that appeared to seal her fate: "There can be 12 white, blue-eyed, blonde men in a room, and they're going to be diverse too because they're going to bring a different life experience and life perspective to the conversation."

The comment seemed to be a clumsy defense of Apple's almost exclusively white male executive leadership and was widely reported. Smith later apologized in an internal email, but a few months later she was gone from the company. It's not clear if she quit or was fired. She had been with the company for twenty years and had received several promotions, rising through the ranks from the head of retail recruitment to become head of diversity, so her sudden departure was a surprise. She was replaced by Christie Smith, a managing principal from the consulting firm Deloitte.

In December 2013, Apple revised its corporate charter to increase boardroom diversity after a complaint by two shareholder groups, Trillium Asset Management LLC and the Sustainability Group. At the time, former Avon CEO Andrea Jung was the only woman on Apple's board. Following the complaint, Apple revised its charter to say, "The nominating committee is committed to actively seeking out highly qualified women and individuals from minority groups to include in the pool from which board nominees are chosen," according to Bloomberg.

The company is also under pressure from an activist shareholder who wants to increase diversity in Apple's executive suite and boardroom. Twice at shareholder meetings, Apple shareholder Antonio Avian Maldonado II has called on Apple to adopt an "accelerated recruitment policy" to increase diversity in senior management and on the board of directors. Maldonado said the company is still too heavily white, and that this will hurt business down the line. "Some of the excuses given by Apple and others—there's not sufficient people in the pipeline, this and that," Maldonado said. "It's bullshit."

At Apple's 2015 shareholders meeting, Maldonado pitched a question directly at Cook about minority leadership at the company. He wasn't satisfied with the answer. "Tim Cook was very defensive, and he presented the two black people on their leadership—but not senior leadership—as a sign of their diversity," Maldonado said. "Personally, I took it as an insult. They were put on the spotlight as 'here's tokenism,' and he didn't seem to accept that."

Apple's board has so far declined Maldonado's proposal. The board said an accelerated recruitment policy "is not necessary or appropriate because we have already demonstrated our commitment to a holistic view of inclusion and diversity and made detailed information about our inclusion and diversity initiatives, and the progress we have made with respect to these initiatives, available on our website at apple.com/diversity."

But Maldonado raises a good point. Apple isn't very diverse, especially at the upper levels, and the board's answer was a disingenuous cop-out. Apple has pushed the issue under the rug. As Maldonado says, the company should have an accelerated recruitment policy—at all levels, and especially the top—because change isn't happening fast enough. "Meaningful change takes time," Apple said in its diversity report. "We're proud of our accomplishments, but we have much more work to do."

Cook's Education Initiatives

Cook has put some initiatives in place to try to increase diversity at all levels. To do this, Apple needs to increase the number of women and underrepresented minorities graduating from college with STEM degrees. Most of the future job growth in the United States will be in technical fields, but only 17.1 percent of industrial engineers in the United States are women. "The reality is," Cook said, "you'll end up having a whole set of jobs that aren't filled. You'll lose talented workforce that should exist. I think it's imperative for the whole country to get behind changing that."

To help counter the dearth of women and minority job candidates, Apple has been entering into several multiyear, multimillion-dollar partnerships with educational nonprofit organizations. Apple launched the Product Integrity Inclusion and Diversity Scholarship, a $10,000 scholarship to help pay for the education of "women, black/African American, Hispanic, or Native American university students." The program debuted in 2014, soon after Apple published its first diversity report, which showed that the company's workforce was dominated by white males worldwide.

Apple is hoping that scholarships will encourage women and minorities to consider a career in tech, and that one day they will have a career in Apple Park.

In 2015, Apple donated more than $50 million to groups that want to get more women, minorities, and veterans working in tech. Apple committed $40 million to the Thurgood Marshall College Fund, which supports students enrolled at public, historically black colleges and universities, known as HBCUs. The schools include Howard University, Grambling State University, and North Carolina A&T State University. The donation will cover scholarships and staff training, and a paid internship program at Apple. Johnny Taylor, president and CEO of the Thurgood Marshall College Fund, said it's the biggest partnership in the fund's history. "What differentiates this partnership with Apple is that it hits on everything that we do—it is the most comprehensive program ever offered to an HBCU organization," he said.

Apple also teamed up with the National Center for Women and Information Technology (NCWIT) to help develop more women technology workers. Apple donated $10 million over four years to NCWIT—the largest single donation the organization has received—to support internships, scholarships, and other education programs, and to reach ten thousand middle school girls over the next few years.

But Apple is not alone. NCWIT is also funded by Microsoft, Google, Symantec, and others in the tech industry, while the Thurgood Marshall College Fund has gotten big donations from the National Basketball Association and Walmart. Both Facebook and Google have partnered with groups to encourage more women to study computer science. Facebook teamed up with Girls Who Code, a computer science early education program; and Google launched an initiative called Made with Code, which teaches 3-D printing and fashion.

"We wanted to create opportunities for minority candidates to get their first job at Apple," said Young Smith, Apple's former head of inclusion and diversity. "There is tremendous upside to that and we are dogged about the fact that we can't innovate without being diverse and inclusive."

Sowing Early Seeds

To increase the pipeline, Apple is reaching even further back than colleges, into high schools, middle schools, and elementary schools. "We're trying to encourage more learning and development in STEM fields because we want to make sure that we've got more of a pipeline coming up through schools starting K through 12," said Apple's VP of HR, Deirdre O'Brien. The issue is personal for Cook, and one reason that education is one of his core values for Apple. As he said in 2015, "I would not be where I am today without a great public education. Too many kids aren't given a good public education. It's not fair."

Apple's commitment to education resulted in a $100 million donation to the U.S. government's ConnectED program in 2014, which earned the company a commendation in President Barack Obama's State of the Union address. Launched under President Obama in June 2013, the ConnectED Initiative is a $10 billion plan to boost broadband connectivity nationwide in K–12 classrooms. When the program launched, less than 40 percent of schools had broadband access, according to the White House. The Obama administration wanted that to be raised to 99 percent by 2018. A 2017 State of the States report from EducationSuperHighway, an advocacy group that tracks broadband connectivity in classrooms, said that 94 percent of public school districts had high-speed Internet, representing nearly 40 million American students. This is a huge achievement, but there's still more to do: 6.5 million students still need high-speed Internet in their schools.

Apple and Cook see the program as a huge success in giving schools across the nation access to technology, support, and important infrastructure. "These kids are born in a digital world, and if they come to school and . . . have an analog environment, it's not conducive to learning," Cook told ABC's Robin Roberts during a tour of ConnectED schools in New York City in 2016. "It's not conducive to creativity. We're bringing digital to the schools here, and we're focused on underserved schools." Cook added, "It's going fabulous. We feel really good."

Apple under Cook has launched its own Everyone Can Code initiative, which offers a comprehensive curriculum to help teach coding to students from Key Stage 1 to university. It includes everything from teacher guides and lessons to coding resources and access to educator forums—all based on Swift, the open-source programming language designed by Apple. Cook believes that coding has become an essential tool for every student, which should be taught in every school around the world.

"If I were a French student and I were ten years old, I think it would be more important for me to learn coding than English," Cook told the French publication *Konbini* during a visit to France in 2017. "I'm not telling people not to learn English, but [coding] is a language that you can use to express yourself to seven billion people in the world. I think that coding should be required in every public school in the world, and we try very hard to make that accessible by making a programming language, that we call Swift, that's as easy to learn as our products are to use. It's the language that everyone needs."

Cook reiterated this during an interview for MSNBC's "Revolution: Apple Changing the World," when he said, "I want America to be strong, first and foremost, and I think to do that, we need to code. It is a language and it is everywhere in our life. It is problem solving. You need critical thinking to know what is fake and what is real." He added, "For Apple, we are taking the responsibility. Businesses should be about more than making revenue and profits."

In addition to the Everyone Can Code initiative, Apple offers free Hour of Code workshops that teach coding in its retail stores. The workshops promise to teach the basics of Swift, whether you're a kid or an adult, an amateur or a budding developer. For those who can't make it to an Apple store, there's the Swift Playgrounds app for iPad, which makes it easy for kids to get started with coding at home or at school. "At Apple we care deeply about education because we love kids and we love teachers," Cook told attendees during an Apple education event at the Lane Tech College Prep High School in Chicago in March 2018 before unveiling the new iPad. "We love creativity and curiosity, and we know that our products can help bring out the creative genius in every kid. That's why education is such a big part of who we are as a company, and has been for forty years." Currently, the iPad is more expensive than the ultra-affordable Chromebooks that dominate classrooms today, but Apple is hoping that education discounts, alongside ClassKit and Schoolwork, a framework and accompanying app that debuted alongside iOS 11.4, will help change that in the coming years.

With ClassKit, developers can create education apps that give students and teachers the ability to connect like never before. The framework enables teachers to discover specific learning activities inside an iPad app, then launch them on students' iPads—inside the necessary apps—with just one tap. It also gives students the ability to privately and securely share progress data with teachers, which can be used for tailored instruction when necessary. The Schoolwork app takes advantage of this framework to let teachers issue assignments inside iPad apps on student devices, then monitor their progress.

Altogether, Apple under Cook has launched a comprehensive set of initiatives, from elementary schools to colleges, to improve the numbers of women and underrepresented minorities going into tech. "We've fundamentally concluded instead of just waiting and going into the four-year school system and seeing how many women and minorities are graduating in coding, which is abysmal, that we had to back up," he said in an interview with *USA Today*. "[We have to go] all the way into elementary school and junior high school in order to fundamentally change the diversity."

Cook said Apple is making progress, especially in recruiting more women. "I suspect that tech as a whole will look dramatically different over a course of time," he said. "It's a wave. . . . The wave has to progress through years, and it will change." Apple's latest Inclusion & Diversity report, published in December 2017, says that female representation at the company is "steadily increasing." The company said 36 percent of its employees under thirty are women, an increase of 5 percent since 2014. Overall, women represent 32 percent of the workforce. Twenty-nine percent of leaders at Apple are women, as are 39 percent of leaders under thirty. The reason Apple breaks out employees under thirty is because this segment represents new blood—the future of the company. As new, younger people come in, the company is slowly becoming more diverse.

Accessibility

Cook has also made accessibility an important focus for Apple. All of Apple's products and software ship with assistive technologies that are designed to ensure that anyone can use them. "We want everyone to enjoy the everyday moments that technology helps make possible," Apple says on its website, "so we work to make every Apple product accessible from the very start."

Every product that Apple makes—the Mac, iPhone, iPad, and Apple Watch—is designed to be used by people with impairments, including the blind and deaf. Apple's VoiceOver technology, for example, which speaks aloud what's displayed onscreen, is one of the features that helps the visually impaired use their devices. It works with iPhone, iPad, Mac, and even Apple Watch.

iOS supports Braille in more than twenty-five languages, and has a Braille keyboard built in. For the hearing-impaired, the iPhone's LED can be set to flash when a call comes in. FaceTime, a free app that makes video calls, is a popular way for deaf users to make calls using sign language. To help the visually impaired, the iPhone's camera app uses facial recognition and VoiceOver to announce when someone is in the shot so that a blind person can take a picture. Owners of the Apple Watch can track workouts in a wheelchair. Live Listen, a feature introduced in iOS 12, turns Apple's AirPods into hearing aids that use an iPhone as a sound-boosting microphone.

"People with disabilities often find themselves in a struggle to have their human dignity acknowledged," Cook said during his acceptance speech for the International Quality of Life Award (IQLA) from his alma mater, Auburn University, in 2013. "They frequently are left in the shadows of technological advancements that are a source of empowerment and attainment for others, but Apple's engineers push back against this unacceptable reality. They go to extraordinary lengths to make our products accessible to people with various disabilities, from blindness to deafness to various muscular disorders."

Cook even acknowledged that making products accessible can be a money loser, but he doesn't care. "We design our products to surprise and delight everyone who uses them, and we never, ever analyze the return on investment," he said. "We do it because it is just and right, and that is what respect for human dignity requires, and it's a part of Apple I'm especially proud of."

Cook frequently highlights Apple's commitment to accessibility. Every summer during Apple's annual WWDC programmers conference, his keynote speech includes accessibility onstage or in a video. In addition, the company highlights accessibility on its App Store, running features on the anniversary of the Americans with Disabilities Act and Autism Awareness Month. Accessibility is front and center on Apple's website, with lots of information for consumers and resources for third-party software developers. For several years, Apple has promoted Global Accessibility Awareness Day during the entire month of May by holding accessibility events, talks, and workshops at its retail stores. Since 2017, Apple has held more than ten thousand accessibility sessions.

To mark Global Accessibility Awareness Day in 2018, Apple announced plans to team up with leading educators for blind and deaf communities across the United States and bring its Everyone Can Code program to their schools. It collaborated with engineers, educators, and programmers from many accessibility communities to make the program as accessible as possible, and vowed to continue working with schools to augment the curricula as needed.

"We're thrilled to kick off the partnership with Apple," said Clark Brooke, superintendent of the California School for the Deaf. "This program is a great way to bring to life the ideas and imagination of our deaf students through coding, while also building a foundation for future careers in software development and technology."

Apple has received several awards for accessibility, including the Helen Keller Achievement Award for VoiceOver from the American Foundation for the Blind. After Apple received the award, it was clear that "accessibility has just become part of the company's DNA," according to Greg Joswiak. "Apple's products are intuitive and accessible right out of the box," said AFB president and CEO Carl R. Augusto. "Apple is truly in a league of its own."

"We see accessibility as a basic human right," Sarah Herrlinger, senior manager for global accessibility policy and initiatives at Apple, told Steve Aquino, a disabled reporter who writes for TechCrunch. According to Aquino, "The accessibility features on iOS are widely regarded as the best in the industry. This is no small feat, one that shouldn't be overlooked, especially if you remember what cell phones were like before the iPhone came along.

"Consider someone with low vision," he said. "He or she may have struggled to use a 'dumbphone' with a display the size of a postage stamp, and a multi-tap keyboard. But then they buy an iPhone, and their whole world changes. . . . Suddenly, they're texting with family and friends, looking up directions, and more with a fluidity like never before. Thus, it isn't hyperbole to say iOS's accessibility features have been every bit as game-changing for the disabled as the iPhone was to the mass phone market." These endorsements and awards show that Apple has made huge strides when it comes to accessibility and that their products are making the world a better, more inclusive place.

Cook reminded Apple's customers of the company's accessibility efforts in October 2016, when it kicked off a major product event (for its lineup of MacBook Pro laptops) with a video titled "Sady," which showcased the unique ways in which people with disabilities use Apple products to learn, communicate, be productive, and enjoy their hobbies. One of the individuals featured in the film was Sady Paulson, a video editor with cerebral palsy, who created the video using Switch Control on a Mac. "That experience was the best thing ever in my life and I won't forget it," Paulson later wrote on her blog. "I am so blessed and thankful for everything that I have. Thank you, Apple and Tim Cook for this amazing opportunity! I do really appreciate everything that you have done with technology for everyone!"

To celebrate Global Accessibility Awareness Day in May 2017, Cook sat down with three YouTubers to discuss the accessibility features built into Apple products—and to explain why the company goes to great lengths to ensure that everyone can use its devices. "Apple is founded on giving people power to create things, to do things that they couldn't do without those tools," Cook told Rikki Poynter, a vlogger and advocate for the deaf. "And we've always viewed accessibility as a human right. And so just like human rights are for everyone, we want our products to be accessible for everyone."

Cook reiterated that the company promotes accessibility not for profit, but to do the right thing. "We feel very strongly that everyone deserves an equal opportunity and equal access," he continued. "So we don't look at this thing from a return on investment point of view. We don't care about that." This isn't a statement Cook is shy about making.

Apple isn't the only company that offers assistive technology—but it is one of the few that makes accessibility a primary focus in everything it does. The values Jobs instilled in Apple long ago have only grown and become even more important with Cook at the helm.

Chapter 11

Robot Cars and the Future of Apple

On August 2, 2018, Apple became the first company in history to reach a market valuation of $1 trillion. It's truly a gigantic number——a one with twelve zeros: $1,000,000,000,000. Apple's stock hit $207.05 just before noon. Most of the gains in its stock price occurred under Tim Cook. Since Cook took over the company, AAPL stock has tripled in value. Some experts attribute the trillion-dollar valuation to the success of the iPhone, the iPhone X especially. Though the iPhone X shipped fewer units than previous iPhones, the new design allowed Apple to raise its price, contributing to significantly higher revenues. Steve Jobs may have given life to the iPhone, but Cook took the product to new heights, and the company has flourished.

The trillion-dollar valuation is a testament to the astonishing growth Apple achieved under Cook's watch. In a memo to employees, he applauded Apple's success and thanked staffers for their hard work. Though he said they should be proud of this accomplishment, he also made it clear that "it's not the most important measure of our success." He emphasized the importance of values, as always, stating that "financial returns are . . . the result of Apple's innovation, putting our products and customers first, and always staying true to our values." It's clear from the memo that he deeply appreciates the contributions of all Apple employees, from entry-level to executive. He ended the memo by crediting Steve Jobs for creating this incredible company and reinforcing the amazing role that Apple products play in the lives of people around the world:

> Steve founded Apple on the belief that the power of human creativity can solve even the biggest challenges—and that the people who are crazy enough to think they can change the world are the ones who do. In today's world, our mission is more important than ever. Our products not only create moments of surprise and delight, they empower people all around the globe to enrich their lives and the lives of others. Just as Steve always did in moments like this, we should all look forward to Apple's bright future and the great work we'll do together.

Future Initiatives

And though we don't know what Apple is planning next, the company's future does indeed look bright. But Apple's success hasn't been without setbacks. It's going to be tough for Cook and Apple to follow the iPhone, perhaps the single most successful product of all time, but if Cook and his lieutenants are looking around for the next sector to disrupt, cars and health care could be high on the list. They are two of the biggest industries on the planet. Health care is the biggest industry in the United States, worth $24.5 billion in 2016, according to the Inc. 5000, an annual list of the fastest-growing private U.S. companies. Logistics and transportation is fourth at $12.8 billion a year. The Apple Watch is on track to be a major new category in health care, but Apple's car project, Project Titan, appears to be stalled, perhaps even moribund.

Project Titan, one of the most ambitious and intriguing developments under Cook's leadership, is a secretive self-driving-car project that has suffered a number of twists and turns. The super-secret project came to light in 2015 when Apple was sued by A123 Systems, an electric car battery maker based in Massachusetts, for purportedly poaching many of its engineers. "Apple is currently developing a large scale battery division to compete in the very same field as A123," the company alleged in its lawsuit, which accused Apple of embarking "on an aggressive campaign to poach" its staff and "raid" its business. Apple was said to be taking so many of its specialized engineers that A123 was forced to close down projects of its own and "scramble to find replacement [staff] at substantial cost."

Billionaire Apple investor Carl Icahn added fuel to the fire several months later when he penned an open letter to Cook, which acknowledged the increasing rumors surrounding an "Apple Car," which was expected to enter the automobile market by 2020. "We believe the rumors," Icahn wrote. "While we respect and admire Apple's predilection for secrecy, the company's aggressive increases in R&D spending . . . have bolstered our confidence that Apple will enter two new product categories: television and car. Combined, these two markets represent $2.2 trillion, three times the size of Apple's existing markets."

Cook reportedly approved Project Titan in 2014 and assigned it to Steve Zadesky, a former Ford engineer who was then working as Apple's vice president of product design. But discussions surrounding an Apple

Car date back to 2008, when Jobs, who had recently introduced the world to the iPhone, started taking an interest in Tesla Motors and its new electric car that was making big waves in the automotive industry. Tony Fadell, former head of the iPod division, was one of the Apple executives who was a part of those discussions.

Fadell believed Apple could build a car, and he compared the design of a motor vehicle to that of a product the company had already mastered. "A car has batteries; it has a computer; it has a motor; and it has mechanical structure. If you look at an iPhone, it has all the same things," Fadell said. Apple seemed primed to enter the automobile industry. "But the hard stuff is really on the connectivity and how cars could be self-driving," he continued, and Jobs ultimately decided not to pursue self-driving cars, in part because the automotive industry was suffering significant hardship at the time. But five years later, Cook saw an opportunity for Apple to shake up the giant auto industry and put another dent in the universe.

Zadesky was given permission to employ up to one thousand people to flesh out the Project Titan team by early 2015, and A123 Systems wasn't the only company Apple was poaching from; designers and engineers from the likes of BMW and Mercedes-Benz had also made the move to Cupertino to be a part of the team building Apple's first car. They began by studying ways in which they could reinvent almost everything in a car, including motorized doors that opened and closed silently, virtual or augmented reality displays, and improved sensor systems that weren't as conspicuous as the arrays of sensors on other self-driving cars. The team even investigated the possibility of reinventing the steering wheel by making it spherical—like a globe—which could allow for better lateral movement.

Apple had its sights set on Tesla talent, too. It was picking up so many former Tesla employees that Tesla CEO Elon Musk once called the Apple

Car project a "Tesla Graveyard." "They have hired people we've fired," Musk told the German newspaper *Handelsblatt* in late 2015. "If you don't make it at Tesla, you go work at Apple." Musk believed a car was "the next logical thing to finally offer a significant innovation" for Apple—but he warned that building a car is very difficult. Apple discovered this the hard way.

When Zadesky left Apple for "personal reasons" in January 2016, after sixteen years with the company, rumors of turmoil within the Project Titan team began to surface. It was reported that Apple Car workers were being asked to meet unattainable deadlines, while the management team was unclear on exactly what they wanted Project Titan to deliver. Zadesky's plan was to build a semi-autonomous car, which would add some robot-driving abilities but would still rely on a human driver, while Jony Ive's industrial design team was pushing for a fully autonomous system that would allow Apple to completely "reimagine the automobile experience." But somehow, what started out as a plan to build its own Apple-branded self-driving car has shifted focus to building systems that would power cars built by other manufacturers.

In July 2016, Apple assigned Project Titan to Bob Mansfield, former senior vice president of Mac hardware engineering, who retired in June 2012 after thirteen years only to rejoin the company four months later, working on "future projects" as the senior vice president of technologies. It was reported in September 2016 that dozens of employees had been laid off as Apple "rebooted" the project in an effort to give it real purpose. More than one hundred employees left a month later, at which point Bloomberg reported that Apple would give its automotive initiative until late 2017 before making a final decision on its fate.

The future of Project Titan looked bleak at this point, and dreams of an Apple Car—or at the very least an Apple-powered car—were fading. It seemed as though CarPlay, the infotainment system based on iOS, which launched alongside iOS 7 in 2014, was as far as Apple's automotive efforts would extend. But like many of the Apple projects that endured rocky starts plagued by teething troubles, Project Titan lived on and grew more promising as it entered 2017.

After securing a permit from the California Department of Motor Vehicles to test self-driving cars on public roads in January 2017, Apple took its own self-driving platform to the streets. It had been baked into a handful of Lexus RX450h SUVs, each outfitted with a plethora of cameras, radars, and sensors—including a high-end Velodyne 64-channel lidar that's designed to help marine vessels and vehicles automatically detect objects and navigate. Cook spoke about Project Titan for the first time in June 2017, when he confirmed to Bloomberg that Apple was "focusing on autonomous systems." "It's a core technology that we view as very important," Cook added. "We sort of see it as the mother of all AI projects. It's probably one of the most difficult AI projects actually to work on."

Apple ramped up its fleet going into 2018, and is now believed to have around forty-five autonomous Lexus SUVs, which can be spotted driving around Silicon Valley. Its self-driving vehicle technology is also being worked into a shuttle service the company has reportedly dubbed "PAIL," an acronym for "Palo Alto to Infinite Loop," which will transport employees between Apple's offices around Silicon Valley. Employees familiar with the project told the *New York Times* that Apple will again use a commercial vehicle retrofitted with its own autonomous technology.

Right now, the status of Apple's Project Titan isn't clear. It may or may not be on track. But many of Apple's biggest projects went through development hell. Apple's retail stores, for example, were scrapped at the last minute and started over. Likewise, nothing went right for the iPhone until the last few months of its development. But Project Titan looks to be a different order of failure. It's not just product development that has failed, but also hiring, management, and perhaps vision. It's "the greatest failure of anything of the Tim Cook era," said analyst Horace Dediu. "Because clearly they put a lot of work into that. . . . They hired a lot of people and clearly there's nothing to show for it."

Apple focuses so much on manufacturing, but this didn't seem to carry over to Project Titan. They wanted to figure out a new way to manufacture cars. Apple likely looked at everything, from alternative body materials—most of the current auto industry relies heavily on stamping sheets of steel—to different purchasing models. And the likelihood is that they couldn't come up with good answers for some, or many, or perhaps all of these questions. The decision went all the way to the board, it's rumored, and the board wasn't convinced there was enough to justify the huge manpower and expenditure to try to disrupt the auto industry.

Cook committed a cardinal sin at a functional organization like Apple: He hired too many outsiders too quickly. If the rumors are true, Apple hired more than a thousand outside auto experts and within a couple of years had laid them off. Apple grew this project way too quickly, instead of doing it more organically. Dediu told the story of Doug Melton, a veteran Apple programmer, who built the Safari web browser in the early 2000s. At the time, Apple was using Microsoft's Internet Explorer browser, and Steve Jobs didn't want to be dependent on Microsoft for such an important app. When Melton was hired to build Safari, he was told he could hire one more person. The two of them built a demo that simulated how Safari would work. After Steve Jobs signed off on it, they were allowed to hire a third person to build it. "It was only allowed to grow the way a start-up grows," said Dediu. "Show that you are making progress." Hopefully Cook has learned from his mistakes, but only time will tell what becomes of Project Titan.

Apple Park

Apple's giant new spaceship HQ was Steve Jobs's last product at Apple, and in some ways, it was one of Cook's first. Apple Park opened in April 2017, even as the campus was still being built. Apple staffers moved into their new offices in small groups as construction workers put the finishing touches on the huge building and its landscaped grounds. Moving in was a big job; it took more than a year. While Apple staffers moved from the old Infinite Loop campus a couple of miles away, Apple was emptying leased buildings all over Silicon Valley and moving workers into newly empty offices at Infinite Loop. It was "a logistical nightmare," said an Apple employee who asked to remain anonymous.

In April 2006, Jobs announced to the city council of Cupertino that Apple had acquired nine contiguous properties to build a second campus. The dream was to have everyone (or as many people as possible) under one roof, something that was already the case on the Disney Pixar campus in Emeryville, which Jobs had a big hand in designing and which was in many ways a prototype for Apple Park.

For the last two years of his life, Jobs dedicated an enormous amount of time to the Apple campus. In June 2011, just a couple months before he stepped down as Apple CEO and four months before he died, he appeared again in front of the Cupertino City Council to ask permission for Apple to build the campus. This twelve-thousand-person campus was to be built on a former Hewlett-Packard property located between North Tantau and North Wolfe avenues, Homestead Road, and the 280 freeway. Today, that address is listed as 1 Apple Park Way.

It was a large stretch of land totaling 175 acres. Jobs's proposal was for an enormous circular building that was quickly nicknamed the "mothership" due to its resemblance to a UFO. "It's a little like a spaceship landed," Jobs said. Years earlier he had joked that the first-generation iMac looked like it was "from another planet, a good planet. A planet with better designers." This building resembled the craft that planet would send to Earth.

"There's not a single straight piece of glass in this building," Jobs said. The campus would be largely free of visible cars, due to an enormous underground parking lot. Doing this would decrease the surface parking by 90 percent—down to just twelve hundred spaces. The new campus would also function as its own dedicated power source, using "natural gas and other ways that are cleaner and cheaper" than relying on the existing power grid. It would include its own auditorium for Apple media events of the kind that Jobs had perfected. (This was later named the Steve Jobs Theater.) It would have a visitors center, a $75 million gym, and a cafeteria big enough to feed three thousand people at the same time—and the entire staff of fourteen thousand people each day. Jobs wasn't being shy about its ambitions. "We do have a shot at building the best office building in the world," he told the council members. "Architecture students will come here to see this," he boasted.

The company's plans were grandiose. The land already had thirty-seven hundred trees planted. Jobs wanted to almost double this to six thousand trees, so he hired an arborist from Stanford. The rolling green design of the land was reportedly inspired by Stanford University's Dish, a hiking area near the campus with rolling hills topped by a giant radio telescope. But it was also landscaped to direct airflow from nearby hills into the building, to aid with natural cooling.

Jobs was as exacting as ever. He worked closely with both the architect

Norman Foster and Apple's design chief, Jony Ive, who temporarily stepped away from his day-to-day job running the Apple industrial design studio to oversee construction of the campus. A profile of Apple Park, written by Steven Levy, noted some of the perfectionist touches, which sounded characteristically Jobsian. Levy described how much attention had been "devoted to giant glass panels, custom-built door handles, and a 100,000-square-foot fitness and wellness center complete with a two-story yoga room covered in stone, from just the right quarry in Kansas, that's been carefully distressed, like a pair of jeans, to make it look like the stone at Jobs' favorite hotel in Yosemite."

Stefan Behling, a Foster partner who became one of the project leads, recalled Jobs's specific demands: "He knew exactly what timber he wanted, but not just 'I like oak' or 'I like maple.' He knew it had to be quarter-cut. It had to be cut in the winter, ideally in January, to have the least amount of sap and sugar content. We were all sitting there, architects with gray hair, going, 'Holy shit!'"

Campus Opening

Originally the plan was to open in 2015, but construction delays meant that this didn't happen until April 2017. The first event took place at the Steve Jobs Theater on September 12, 2017, for the unveiling of the iPhone X and iPhone 8, signaling the tenth anniversary of Apple's most popular product. Cook opened the event with an emotional tribute to Jobs. "Steve means so much to me and to all of us," he said. "There's not a day that goes by I don't think about him."

The Apple Park name was first announced to the public in February 2017. It was a simple name, like Apple Watch, moving away from the "i" prefix that had accompanied the Jobs-era naming patterns. "Maybe it should be called the Steve Jobs Campus?" Stephen Fry had remarked when visiting the site during construction. "Oh, Steve made his views on that very clear," said Cook. This summed up much of the relationship between the Cook and Jobs eras at Apple. Jobs was still Apple's guiding light, his perfectionist principles even dictating what the new site should *not* be called. But it was also Cook's Apple now.

Not Everything Was a Success

Not everything was straightforward about the new HQ. Within weeks, there were reports that local emergency services had been called out after employees who failed to see the crystal-clear glass had cut their heads walking headlong into the glass walls and doors.

Not everyone was enamored with the new HQ, either. The Apple Park wasn't accessible to anyone but Apple staffers. A tall fence kept anyone from wandering up to it. An op-ed published in Wired criticized the structure for being purposely distanced from the public. "The best, smartest designers and architects in the world could have tried something new," the article noted. "Instead it produced a building roughly the shape of a navel, and then gazed into it."

Alissa Walker, a writer and editor for Curbed, the interior design and architecture site, also criticized its isolation. "Is Apple going to make the grounds open to the public so they can enjoy the fifty billion trees that he'll be planting?" she wrote. "Will there be any kind of programming in the new auditorium that can expose the next generation to careers in technology and science? Could you share your awesome private transit system with the public?"

Allison Arieff, a prominent architecture and design writer, criticized the campus for being isolated from both housing and public transit. She said its location was guaranteed to make for miserable commutes in an area already famous for its miserable commutes. "Building campuses on isolated suburban tracts guarantees long commutes, and this is one of the worst in the country," she wrote. She noted that the building has as much space devoted to parking spaces as offices, and no childcare facilities. She likened it to the suburban office parks of the 1950s, and wondered why Apple hadn't been more forward-thinking. "In a region where the disruption of existing norms is everything, why does this decades-old paradigm of the office persist?"

When I visited the Park in March 2018 for a series of executive interviews, I found the building impressive but sterile. Like many of Apple's big retail stores, the building itself is imposing, but the uniformity of wood and stone makes it lifeless. Everything inside is built to the same specification—all the tables, chairs, stools, and coffee bars are the same. All the work pods have the same layout and office furniture. It's uniform on a gigantic scale, and there's nothing quirky or human about it. It's a giant, perfect cathedral of concrete and glass: exacting in every detail but devoid of humanity.

Fostering Collaboration

The concept wasn't always for Apple Park's main building to be perfectly round. Before they settled on the circular campus, Jobs reportedly insisted that it should look more like a big cloverleaf. When he shared the drawings with his family, it took his son Reed to point out that this would make Apple Park looked like an enormous penis from the air. "You're never going to be able to erase that vision from your mind," Jobs said when he reported the news back to his architectural team.

The circular idea made sense, though. There was a purity to it. The

2.8-million-square-foot circular building would allow Apple to break records for the largest pieces of curved glass in a building. But it would also foster collaboration. Mashable editor Lance Ulanoff questioned Phil Schiller about this, asking what "would happen if one team was seated on one side of the circle and another all the way on the other side? Would collaboration suffer?" Schiller immediately corrected him. "Quite the opposite," he said. "The design of the new campus has been all about fostering collaboration [because] the internal and external surface of the ring are the hallways, and they completely traverse the space. So you can walk through the entire space, both on the inside and outside perimeter and go from section to section."

Most would associate this openness more with Cook's Apple, though it contains traces of Steve Jobs. True, Jobs liked to have siloed teams, off working on their own projects. But he had also begun to embrace a more collaborative approach while at Pixar later in his career. Collaboration and separation were two of his competing impulses.

In his book *Creativity Inc.*, Pixar's Ed Catmull recalls how these two impulses conflicted when Jobs was working on designing the Pixar HQ years earlier. "Steve's first pass at a design was based on some peculiar ideas he had about how to force interaction among people," he wrote. Jobs had proposed single women's and men's bathrooms located off the main atrium. He hoped that by restricting the bathrooms, staffers would be forced to walk to the center of the building, and would interact with co-workers along the way. But the plan didn't go down well. Pixar staffers complained at an offsite meeting where Jobs presented his plan, and he reluctantly shelved it.

But when Jobs and Catmull visited Disney's offices, Jobs "saw firsthand the way that the Disney people took advantage of the open floor plan, sharing information and brainstorming. Steve was a big believer in the power of accidental mingling; he knew that creativity was not a solitary endeavor."

Returning to Pixar, Jobs met with the architects and developed a plan for a single building with no "barriers" to prevent staff interaction. The stairs were "open and inviting." They would encourage staff to see and greet each other as they entered and left. Most of the building's shared spaces—the toilets, conference rooms, mailroom, screening theaters, and eating areas—were located centrally near the atrium. "Everything about the place was designed to encourage people to mingle, meet, and communicate," Catmull wrote. "People encountered each other all day long, inadvertently, which meant a better flow of communication and increased the possibility of chance encounters. You felt the energy in the building."

It Seems to Be Working

The same seems to be true at Apple Park. Though the building is closed to the public, I was able to visit a couple of times in March 2018, and the building was a hive of activity. All over campus, Apple staffers were walking through the grounds or around the inner or outer walkways. All the corridors and building atriums are dotted with coffee stations, as well as chairs and tables for impromptu meetings. Many were occupied with people chatting. It was impossible to tell what they were chatting about, of course, but it looked like they were doing what Jobs had hoped: meeting and interacting.

Greg Joswiak, the VP of worldwide product marketing, agrees. He thinks Jobs's plan to foster more meetings and collaboration through the building's architecture is working. There are workers meeting at the stations around campus all the time. "Sometimes, we're having these little impromptu meetings that we can get something done in ten or fifteen minutes sitting there that didn't have to require us to have a whole hour meeting with people to get through." He says he loves the open floor plan, because when he has to travel "across the campus, there's a pretty good chance I'm going to see someone along the way. And I give myself a little extra time because we're probably going to stop and have a chat." It's a cultural shift that is "certainly changing the way we do day-to-day work. . . . It's good to see it's paying off." The new innovative headquarters is starting to change Apple for the better.

Enter the X, the Next-Gen iPhone

At just after 10 a.m. on September 12, 2017, Tim Cook kicked off Apple's very first press event at the Steve Jobs Theater in the newly opened Apple Park. Every one of the underground auditorium's one thousand seats was filled by select Apple employees and members of the media who were lucky enough to have received an invite, and they were all awaiting one thing: their first official glimpse of Apple's tenth anniversary iPhone.

"The first iPhone revolutionized a decade of technology and changed the world in the process," Cook reminded attendees. "Ten years later, it is only fitting that we are here, in this place, on this day, to reveal a product that will set the path for technology for the next decade." That product was, of course, the iPhone X, which Apple dubbed "the future of the smartphone."

The iPhone X was special, not just because it marked a decade of the iPhone, but because it did away with a design language iPhone fans had been accustomed to since the original to usher in new technologies. It was the first iPhone with an edge-to-edge Super Retina display and Face ID, and Cook called it "the biggest leap forward since the original iPhone."

Cook invited Phil Schiller up to the stage to gush about the iPhone X's glass and stainless steel design, and its sharp OLED screen—a first for the iPhone—which offered HDR support and True Tone. Schiller, like Cook, seemed excited to show the audience Apple's most impressive smartphone to date, but Apple fans and critics weren't quite as enthusiastic about the dramatic changes that had been made.

Many were upset that the iPhone X's edge-to-edge screen left no room for a physical Home button with a Touch ID fingerprint scanner, while others felt that its "notch," which sat at the top of the display and housed the phone's front-facing camera, speaker, and Face ID sensors, was ugly and obtrusive. Even those who welcomed Apple's advancements found it difficult to stomach the iPhone X's eye-watering price tag, which started at $999 and rose to $1,149 with additional storage.

Cook was happy to defend Apple's decision to charge a premium price—more than it had ever charged for an iPhone. "In terms of the way we price, we price to sort of the value that we're providing," he explained during an Apple earnings call shortly before the iPhone X went on sale. "We're just trying to price it for what we're delivering. And iPhone X has a lot of great new technologies in there that are leading the industry, and it is a fabulous product."

Analysts weren't convinced that Apple's brave move would pay off. The majority predicted the iPhone X would flop, and that Apple would be forced to cut its prices the following year. Some suggested that iPhone X production would cease prematurely because Apple simply wouldn't sell enough units after early demand from its biggest fans had died down. The reality was very different.

Orders for the iPhone X started "very strong," Cook revealed in November 2017, and they remained that way. It outsold every other smartphone in Apple's lineup quarter after quarter, making 2017's iPhone cycle "the first cycle in which the top-of-the-line iPhone model has also been the most popular," Cook proudly stated during an earnings call in May 2018, before comparing its performance to that of a successful football team. "It's one of those things where a team wins the Super Bowl. Maybe you want them to win by a few more points, but it's a Super Bowl winner and that's how we feel about it," he said. "I could not be prouder of the product."

The iPhone X proved that under Tim Cook, Apple could still innovate and charge premium prices.

Chapter 12

Apple's Best CEO?

Jobs was a unique CEO, the likes of which we will probably never see again. He wasn't just the CEO, he was also Apple's chief product officer—the person who made the key decisions about products. Cook hasn't taken on that role, but that's okay, because he doesn't need to. Lots of people expected him to fail because he wasn't "a product guy," says analyst Horace Dediu. "But that's not what he should be."

Unlike at Apple, at most companies, the product guy isn't usually the CEO. They are usually further down in the hierarchy—a designer or engineer, perhaps. The product guys are often somewhat interchangeable. Of course, occasionally there are great ones, like Jony Ive. Jony, now Apple's chief design officer, worked very closely with Steve Jobs over the years. Some onlookers wonder if he still needs Jobs's feedback and ideas and help to continue to design great products, but based on the string of products out of Apple since Jobs's passing, it appears not. Apple continues to innovate with unique products long after his death.

But what many may not realize is that what matters most at a mature company like Apple is not the products but rather the logistics—an efficient supply chain, distribution, finance, and marketing. And Cook has proven his talents for all of these. As a result, according to Dediu, he is the best CEO Apple has ever had.

Dediu is well aware that this sounds heretical. How could Cook be a better CEO than Steve Jobs? Jobs has been deified. He's untouchable, and almost everyone would argue he's the best CEO Apple ever had, by far. He started the company and saved it. He's responsible for some of the biggest breakthroughs in the technology industry, from the first PC (the Apple II) to the first easy-to-use PC for everyone (the Mac), and then the iPod, the iPhone, and the iPad, and tons more.

But "Steve Jobs was never really a CEO," Dediu explains. In fact, he thinks Jobs was a terrible CEO. "He was always the head of product." For a big part of his career, he was arguably a terrible CEO. He succeeded in spite of himself. He was all over the place at Apple when it first started, and the company only survived because there were other people in charge. When he returned to Apple, he was great, but the company was much smaller, and he was in crisis mode. When it settled down, he largely turned over the running of the company to Cook so he could concentrate on doing what he loved best—creating new products with Jony Ive. So Cook was already largely the CEO when Jobs was around, and continued in the same role after he passed. And Cook is well suited to running Apple in ways that Jobs was not. "When you become a giant company with a lot of people in operations and a multifaceted business model," Dediu says, "you need a much more generalist CEO. And that's what Tim Cook always was. . . . [He's] the right guy for the job."

The Apple rank and file has confidence in Cook as well. "We still think our future is very bright," Joswiak says. "We have a lot of great stuff in development. That hasn't slowed down" since Cook took over. "Apple employees are very confident in Tim's leadership. . . . Everywhere I go, people are pretty amazed by Tim."

Can Cook Innovate?

Cook may have the support of the Apple workforce behind him, but there's still a big question hanging over his tenure: Can Apple still innovate like it did under Jobs? Jobs had a spectacular track record. At the very beginning of his career, he helped usher in the era of the personal computer with the Apple II. He followed with the original Macintosh, and then eventually the iPod, iPhone, iPad, and multiple innovations in software, such as Mac OS X, iTunes, and the App Store.

But when Jobs was CEO, he wasn't as venerated as he is now. Back then, Apple was struggling in the PC market. Most pundits urged the company to adopt Microsoft's model of licensing its software to other PC makers, which would likely have been the death of the company. The iPod was seen as a one-off lucky hit (and pundits again urged Apple to license iTunes to other companies at the time). The iPhone was initially derided as an expensive flop. Then-Microsoft CEO Steve Ballmer infamously said, "There's no chance that the iPhone is going to get any significant market share." Steve Jobs certainly proved him wrong.

But it wasn't until the iPhone took off, which was also around the time when Jobs got sick from cancer, that his reputation took a turn. While he was alive, people worried about his ability to innovate. "People forget that," says Joswiak. "We went a long time sometimes between innovations, if you will, between category-changing products."

In fact, if you look at Jobs's career, there are periods of many years between these category-changing products. The Apple II launched in 1977. The first Mac followed seven years later in 1984. After his return to Apple, the first iMac launched in 1998, fourteen years later. The iPod and Mac OS X followed in 2001, three years after the iMac, and the iPhone, launched in 2007, came six years after the iPod. The iPad launched in 2010, three years after the iPhone.

And many of these big products took awhile before they became successful. The iPod didn't sell in big numbers until three years after its launch, when Apple added USB and made it Windows-friendly. The iPhone didn't sell until about three years after its initial launch. Very few Apple products have actually been instant hits out of the gate. Steve Jobs gets credit for these amazing products now, as he should, but it's important to remember that it wasn't always easy for him. Cook faces many of the same struggles, with some innovative products taking awhile to catch on.

Innovations Take Time

In the Cook era, the Apple Watch is following a similar pattern. The first major new product category launched under Cook, the Watch was initially greeted with skepticism and even scorn. Early reviews pegged it as a nice toy but not a world-changing product. But three years later, the Apple Watch is the biggest smartwatch on the market and is bigger than the entire Swiss watch industry. Apple is estimated to have sold forty-six million to date. It is likely to develop significantly in coming years. The Apple Watch is a platform for Apple's ambitions in health and wellness. With software initiatives like HealthKit and ResearchKit, Apple is laying the foundation for a wrist-worn computer that helps wearers monitor and improve their health and fitness. Apple is also rumored to be adding new sensors, possibly for blood glucose monitoring. This will be especially helpful for individuals with diabetes, but also useful for everyone else who wants to see what a meal or a donut does to their blood sugar. Dieting will never be the same.

Aside from big, world-dominating products, Apple under Cook has innovated in many other areas. Apple's AirPods are a giant hit, and they are remapping the wireless headphone space. Apple Pay is slowly taking off and is tipped to become the biggest contactless payment system in the United States; it's projected to account for a third of all payments by 2022. Cook is a big proponent of augmented reality as well. It's early days, but some in the space predict augmented reality will change how we use our devices to interact with the world and will likely be bigger than apps. Face ID, the facial recognition system used in the iPhone X, has been well received for making security easy and painless.

Even something as simple as using an Apple Watch to unlock your Mac, which is surprisingly complex behind the scenes, is a small but telling example of innovation in the Cook era. Like Cook himself, these improvements aren't trumpeted as big breakthroughs, but they add up to a better experience and are leading the rest of the tech industry. Indeed, many may not realize that this is the way Apple has always operated; the big breakthroughs are rare, but smaller incremental improvements are common, and sometimes they add up to big new breakthrough products.

Innovation is always on everyone's minds at Apple, and the staffers sometimes joke that the walls are lined with posters that say, "Have you innovated today?" But "innovation isn't something that you just put a sign up and say people can do it," Joswiak says. "We respect great ideas wherever they come from because great ideas don't always come from us, the senior leaders. Great ideas come from brilliant engineers that may be buried a couple levels deep in your organization. . . . We try to listen to that." Cook has made it clear that he values innovation, at every level. And he has a good eye for new technologies himself. "He's got a good knack for knowing what kind of things ultimately can be great," Joswiak says. "I think his track record has been pretty good there."

Lessons Learned

On paper, Cook is an unlikely candidate to be the leading activist CEO in America. A son of the South from a deep red state, he is white and working class. He studied no-nonsense subjects like engineering and business and worked for decades to make the trains run on time. Early in his career, managing inventory, he built a reputation as somewhat cold and ruthless. He cut scrappy deals with suppliers and seemingly cared only about the bottom line. His predecessor appeared far more liberal. Steve Jobs, a scion of moderate California, had long hair, dressed like a hippie, dated rock stars, and was a vegetarian. He loved Bob Dylan and lived on a commune.

Jobs would have been a more likely candidate to turn Apple into one of the country's biggest progressive companies. But there's the rub. Apple under Jobs always had a reputation as a liberal company, but it wasn't particularly liberal in its actions. It was a Fortune 500 killing machine. It dodged taxes, made no visible charitable donations, and exploited and poisoned workers in Asian countries. Jobs was unapologetic about these behaviors, believing that Apple's contributions existed primarily in the form of its products.

Apple under Cook is different. Cook has proven that he is an ethical man, and his values have become an integral part of the company's operation. He is pushing Apple and the entire tech industry forward, creating an ethical transformation. "Slowly, brands are waking up to the fact that strong ethics and core values are no longer a 'nice to have,' but a necessity," wrote Patrick Quinlan, cofounder and CEO of Convercent, a tech company that sells an ethics and compliance management platform, in an op-ed for Recode. "The internet stripped away barriers between consumers and brands, meaning that transparency and attention to ethics and values is at an all-time high. Brands have to get on board, now. Consider some oft-cited casualties of the digital transformation: Blockbuster, Kodak, and Sears. That same fate awaits companies that can't or won't prioritize ethics and values." While Apple has been accused of ethical lapses—worker exploitation, tax dodging, and planned product obsolescence, to name a few—its stances on issues like privacy and the environment are in contrast to its rivals in Silicon Valley and elsewhere.

Tim Cook has placed a clear emphasis on environmental efforts at Apple. While the Trump administration is stepping back from acting on climate change, companies like Apple are leading the charge. Apple has made world-changing investments in renewable energy, responsible forestry, and sustainable manufacturing. When data centers suck down as much power as medium-sized towns, Apple's efforts to build out solar and wind power are incredibly significant. Apple's operations now run on 100 percent renewable energy in twenty-five countries, and it's starting to bring the supply chain along. If Lisa Jackson's estimates are accurate, Apple's supply chain—which accounts for 70 percent of its carbon footprint—will be 100 percent renewable in a decade or less. And if other manufacturers follow Cook's example, the entire manufacturing industry would likely switch to renewable power as well.

A closed-loop supply chain is also another one of Cook's bold ideas. It has yet to be proven successful, but it's a worthwhile idea—to make new products out of old ones, and ensure that as few resources as possible are extracted from the earth. It's an idea that has a long history. Environmentalists and designers have for decades been calling for cradle-to-grave manufacturing, but to have a company the size of Apple put its weight behind it is a big, important step for the entire industry and the world.

Cook has improved worker conditions in the supply chain as well, but to a lesser extent. Abuses are still widespread, but it's important that a company the size of Apple is exporting its values. Cook has made it clear that this is a priority for him, and other companies are taking note. According to Apple, more than 11.7 million workers have been trained to understand their rights as employees, health and safety regulations, and Apple's Code of Conduct. Apple has spent a lot of money developing advanced manufacturing techniques and processes. It's only right that the same inventiveness should be extended to the factory workers.

Cook has been similarly inventive in his stand on privacy and security. He has said that privacy is a fundamental human right, right up there with freedom of speech and other civic rights. But he's one of the lone voices on this issue, and he's in opposition to most of the rest of Silicon Valley. It's not often acknowledged, but the main business model of Silicon Valley is actually not gadgets and gizmos, but advertising, and in the digital age it is more intrusive than ever. The leaders of Facebook and Google are heavily reliant on encouraging their customers to share more and more data. Apple does not. As a result, Apple may be lagging in AI and other technologies that require users' personal data, but Cook has taken a long-term stance that will serve the company and its customers better in the long run. Because of Cook's values, Apple will likely never experience privacy scandals to the same extent that Facebook has. In March 2018, those scandals wiped $100 billion from Facebook's market cap and saw Mark Zuckerberg hauled before Congress.

Accessibility along with diversity and inclusion are two sides of the same coin. Cook showed his commitment to accessibility by adding it to Lisa Jackson's portfolio, elevating accessibility (and education initiatives) to the highest levels of Apple's management. As a result, Apple's products have received high praise from accessibility advocates. In 2017, the company won three major awards for innovations in accessibility. Being blind shouldn't be a barrier to using the iPhone, and Apple is working hard to ensure that its products are for everyone.

Cook's commitment to inclusion and diversity comes from his experience growing up and living as a gay man in a southern state. Coming out was a brave act of civil duty. Cook is likely the most private man at the world's most visible company, yet he sacrificed a portion of his own privacy for the greater good. By coming out publicly, he allowed a lot of marginalized people to take courage in their identities. He helped to further normalize gay people, showing that a gay man can quite competently run the world's biggest company. And he has created initiatives to try to ensure the widest possible source of talent for Apple's workforce. He is right when he says the best companies in America are the most diverse, and Apple is on its way to having a more diverse workforce. Progress is slow, but it's encouraging to hear that in 2017, half of Apple's new hires in the United States were from underrepresented groups in tech.

Cook is proving the adage that it's possible to do good while also doing well. Steve Jobs once said that companies were mankind's best invention for getting groups of people to pull in the same direction. Cook is taking that one step further. As he has said, "I don't think business should only deal in commercial things. Business to me is nothing more than a collection of people. If people should have values, then by extension a company should have values." And though Apple has become the world's first trillion-dollar company under his leadership, Cook has done so much more. He has made Apple a better company and the world a better place.

Acknowledgments

I'd like to thank my wife, Traci, and our kids for their support and encouragement, and for patiently putting up with my many absences during evenings and weekends.

I'd also like to thank my literary agent, Ted Weinstein, and the edit team at Portfolio / Penguin Random House, especially Stephanie Frerich, Niki Papadopoulos, and Rebecca Shoenthal. They deserve big thanks for their great job shepherding the book from concept to completion.

I wouldn't have been able to write the book without my colleagues from the *Cult of Mac* blog. In particular, huge thanks go to Killian Bell and Luke Dormehl, who provided invaluable help with writing and research. Kudos also to Lewis Wallace, Buster Heine, Ed Hardy, Charlie Sorrel, Stephen Smith, David Pierini, Graham Bower, Ian Fuchs, Ami Icanberry, and Erfon Elijah, for running the blog and the *CultCast* podcast in my many long absences. Thanks also to Natalie Jones for helping with research and interviews.

I'm very grateful to Steve Dowling and Fred Sainz from Apple PR, who provided invaluable help and assistance. I also need to thank the Apple executives who agreed to talk with me about Apple and Tim Cook: Greg Joswiak, Lisa Jackson, Deidre O'Brien, and Bruce Sewell, as well as a couple of others who asked to remain anonymous.

Thank you also to all the interview subjects who graciously devoted their time to talking about Apple and Tim Cook. The book benefited greatly from the reporting of others, especially Yukari Kane (*Haunted Empire*); Adam Lashinky (*Inside Apple* and several features for *Fortune*); and Brent Schlender and Rick Tetzeli (*Becoming Steve Jobs*).

Notes

Introduction: Killing It

x **Its stock has nearly tripled:** Anita Balakrishnan and Sara Salinas, "Apple's Cash Hoard Falls to $267.2 Billion," CNBC, May 2, 2018, accessed September 10, 2018, www.cnbc.com/2018/05/01/apple-q2-2018-earnings-heres-how-much-money -apple-has.html.

x **despite its spending nearly $220 billion:** Stephen Grocer, "Apple's Stock Buybacks Continue to Break Records," *New York Times,* August 1, 2018, accessed September 10, 2018, www.nytimes.com/2018/08/01/business/dealbook/apple-stock-buybacks .html.

x **For perspective, the U.S. government:** *Financial Report of the United States Government, FY 2017* (Washington, DC: Federal Accounting Standards Advisory Board, 2017), 10, www.fiscal.treasury.gov/fsreports/rpt/finrep/fr/17frusg/02142018_FR(Final) .pdf.

x **Q1 of 2018:** Apple Inc., "Apple Reports First Quarter Results," news release, February 1, 2018, www.apple.com/ca/newsroom/2018/02/apple-reports-first-quarter -results.

x **By comparison, Facebook:** "Facebook Reports Fourth Quarter and Full Year 2017 Results," Facebook, January 31, 2018, accessed September 10, 2018, https://investor .fb.com/investor-news/press-release-details/2018/facebook-reports-fourth-quarter -and-full-year-2017-results/default.aspx.

x **Not to mention:** "Microsoft Annual Report 2017," Microsoft Store, accessed September 10, 2018, www.microsoft.com/investor/reports/ar17/index.html.

x **Apple has sold:** "Apple iPhone Sales 2018," Statista, accessed September 10, 2018, www.statista.com/statistics/263401/global-apple-iphone-sales-since-3rd-quarter-2007/.

x **While Android may:** Chuck Jones, "Apple Continues to Dominate the Smartphone Profit Pool," *Forbes*, March 3, 2018, accessed September 10, 2018, https://www.forbes.com/sites/chuckjones/2018/03/02/apple-continues-to-dominate-the-smartphone-profit-pool/#492b439361bb.

x **While Apple sells:** "Apple Inc Gross Profit Margin (Quarterly)," YCharts, accessed September 10, 2018, https://ycharts.com/companies/AAPL/gross_profit_margin.

x **Apple's market share:** "iPhone Will Grab More Market Share as Samsung Falls in 2018," Cult of Mac, February 13, 2018, accessed September 10, 2018, www.cultofmac.com/528725/iphone-will-grab-market-share-samsung-falls-2018.

x **Although computers play:** "Apple's Market Share Increases After Mac Shipments Rise in 2017," Cult of Mac, January 12, 2018, accessed September 10, 2018, www.cultofmac.com/523037/apples-market-share-increases-mac-shipments-rise-2017/.

xi **But since Cook:** Mark Rogowsky, "Race to $1 Trillion: Tim Cook, Apple Redefining 'Winner Take All,'" *Forbes*, August 3, 2017, accessed September 10, 2018, www.forbes.com/sites/markrogowsky/2017/08/03/apple-is-redefining-winner-take-all-as-the-cook-era-hits-new-peak/#26981cc44391.

xi **the Apple Watch:** Horace Dediu, Twitter post, May 2, 2018, 4:46 a.m., https://twitter.com/asymco/status/991645023119790080.

xi **quarter over quarter:** Todd Haselton and Anita Balakrishnan, "Apple Watch Sales Up 50 Percent for Third Consecutive Quarter," CNBC, November 2, 2017, accessed September 10, 2018, www.cnbc.com/2017/11/02/apple-watch-sales-up-50-percent-for-the-third-quarter-in-a-row.html.

xi **Apple's watch business:** "Apple Watch Made $1.5 Billion More Than Rolex Last Year," Cult of Mac, April 26, 2016, accessed September 10, 2018, www.cultofmac.com/425038/apple-watch-made-1-5-billion-more-than-rolex-last-year/.

xi **Apple's AirPods are:** Joe Rossignol, "KGI: AirPods Shipments Will Double Next Year Given Strong Demand," MacRumors, accessed September 10, 2018, www.macrumors.com/2017/12/04/kuo-airpods-shipments-double-in-2018.

xi **With the new HomePod:** Felix Richter, "Infographic: Apple's 'Other Products' on the Rise," Statista, June 26, 2018, accessed September 10, 2018, www.statista.com/chart/14433/apples-other-products-revenue/.

xi **Responsible for $9.1 billion:** Apple Inc., *Q2 2018 Unaudited Summary Data*, www.apple.com/newsroom/pdfs/Q2_FY18_Data_Summary.pdf.

xi **Fortune 500 company:** Mike Murphy, "Apple's iPhone Business Has More Revenue Than Amazon," Quartz, November 2, 2017, accessed September 10, 2018, https://qz.com/1119147/apple-is-two-fortune-100-businesses-and-three-fortune-250-businesses-in-one-aapl.

xi **Some pundits see:** Chloe Aiello, "Apple Services Revenue Could Soar to About $50 Billion Faster Than Even CEO Tim Cook Lets On: Tech Investor Calacanis," CNBC, August 1, 2018, accessed September 10, 2018, www.cnbc.com/2018/08/01/apple -services-will-be-a-money-printing-machine-jason-calacanis.html.

xii **In late 2017:** Apple Inc., "SEC Filings," Annual Report, accessed September 10, 2018, http://investor.apple.com/secfiling.cfm?filingid=1193125-17-380130&cik=320193.

xii **Apple believes accessibility:** "Accessibility," Apple, accessed September 10, 2018, www.apple.com/accessibility/.

xii **Apple believes education:** "Education," Apple, accessed September 10, 2018, www .apple.com/education/.

xii **Apple drives environmental:** "Environment," Apple, accessed September 10, 2018, www.apple.com/environment/.

xiii **Apple believes diverse:** "Inclusion & Diversity," Apple, accessed September 10, 2018, www.apple.com/diversity/.

xiii **Apple believes privacy:** "Privacy," Apple, accessed September 10, 2018, www.apple .com/privacy/.

xiii **Apple educates and:** "Supplier Responsibility," Apple, accessed September 10, 2018, www.apple.com/supplier-responsibility.

Chapter 1: The Death of Steve Jobs

1 **Cook, surprised by:** Brent Schlender and Rick Tetzeli, *Becoming Steve Jobs: The Evolution of a Reckless Upstart into a Visionary Leader* (Toronto: Signal, 2016), 404.

2 **"There'll be more":** "Biographer Isaacson Describes the Man Who Co-founded Apple . . . ," Commonwealth Club, December 14, 2011, accessed September 10, 2018, www.commonwealthclub.org/events/archive/transcript/walter-isaacson-talks -steve-jobs.

2 **"You can bet":** David Pogue, "Steve Jobs Reshaped Industries," *New York Times,* August 25, 2011, accessed September 10, 2018, https://pogue.blogs.nytimes.com /2011/08/25/steve-jobs-reshaped-industries/.

2 **He was prepared:** Schlender and Tetzeli, *Becoming Steve Jobs,* 404.

3 **"over [to his house]":** Schlender and Tetzeli, *Becoming Steve Jobs,* 404–5.

3 **"unfortunately, it didn't":** Schlender and Tetzeli, *Becoming Steve Jobs,* 405.

4 **Another possible candidate:** Philip Elmer-Dewitt, "Scott Forstall Is Apple's 'CEO-in-Waiting' Says New Book," *Fortune,* January 17, 2012, accessed September 10, 2018, http://fortune.com/2012/01/17/scott-forstall-is-apples-ceo-in-waiting-says-new -book/.

4 **a "mini-Steve":** Adam Satariano, Peter Burrows, and Brad Stone, "Scott Forstall, the Sorcerer's Apprentice at Apple," *Bloomberg Businessweek,* October 13, 2011, accessed September 10, 2018, www.bloomberg.com/news/articles/2011-10-12/scott-forstall -the-sorcerers-apprentice-at-apple.

5 **"Nobody would make":** Adam Lashinsky, "Apple's Tim Cook: The Genius Behind Steve Jobs," *Fortune,* November 24, 2008, accessed September 10, 2018, http://fortune.com/2008/11/24/apple-the-genius-behind-steve/.

5 **The other half:** SEC Form 4, Statement of Changes in Beneficial Ownership, accessed September 10, 2018, www.sec.gov/Archives/edgar/data/320193/000118143111047180/xslF345X03/rrd320669.xml.

6 **"The Board has complete":** "Steve Jobs Resigns as CEO of Apple," Apple, August 24, 2011, accessed September 10, 2018, www.apple.com/newsroom/2011/08/24Steve-Jobs-Resigns-as-CEO-of-Apple.

6 **The same day:** Yukari Iwatani Kane, "Jobs Quits as Apple CEO," *Wall Street Journal,* August 25, 2011, accessed September 10, 2018, https://www.wsj.com/articles/SB10001424053111904875404576528981250892702.

6 **AllThingsD's Walt Mossberg:** Walt Mossberg, "Essay: Jobs's Departure as CEO of Apple Is the End of an Extraordinary Era," AllThingsD, August 24, 2011, accessed September 10, 2018, http://allthingsd.com/20110824/jobs-leave-a-legacy-of-changed-industries.

6 **People looked for clues:** Jordan Golson, "Steve Jobs to Remain on Disney Board," MacRumors, August 24, 2011, accessed September 10, 2018, www.macrumors.com/2011/08/24/steve-jobs-to-remain-on-disney-board.

7 **a "sudden worsening":** Adam Satariano, "Apple's Jobs Resigns as CEO, Will Be Succeeded by Tim Cook," Bloomberg, August 25, 2011, accessed September 10, 2018, www.bloomberg.com/news/articles/2011-08-24/apple-ceo-steve-jobs-resigns.

7 **Jobs had left:** David Gardner and Ted Thornhill, "Steve Jobs Dead: Apple Boss Left Plans for 4 Years of New Products," *Daily Mail,* October 8, 2011, accessed September 10, 2018, www.dailymail.co.uk/news/article-2046397/Steve-Jobs-dead-Apple-boss-left-plans-4-years-new-products.html.

8 **"I am looking forward":** Jacqui Cheng, "Exclusive: Tim Cook E-mails Apple Employees: 'Apple Is Not Going to Change,'" Ars Technica, August 25, 2011, accessed September 10, 2018, https://arstechnica.com/gadgets/2011/08/tim-cook-e-mail-to-apple-employees-apple-is-not-going-to-change.

8 **Cook continued this tradition:** "Like Steve Jobs, Apple CEO Tim Cook Also Responds to His Email," Cult of Mac, July 28, 2015, accessed September 10, 2018, www.cultofmac.com/111374/like-steve-jobs-apple-ceo-tim-cook-also-responds-to-his-email.

8 **"Tim, just wanted":** Eric Slivka, "A Look at Apple's Handling of Customer Emails to Executives as Tim Cook Takes Charge," MacRumors, August 30, 2011, accessed September 10, 2018, www.macrumors.com/2011/08/30/a-look-at-apples-handling-of-customer-emails-to-executives.

9 **He had defied:** "Prognosis," Hirshberg Foundation for Pancreatic Cancer Research, accessed September 10, 2018, http://pancreatic.org/pancreatic-cancer/about-the-pancreas/prognosis.

9 **The 4S's big new feature:** Luke Dormehl, *Thinking Machines: The Quest for Arti-*

ficial Intelligence—and Where It's Taking Us Next (New York: TarcherPerigee, 2017), 102.

10 **"among the greatest":** Kori Schulman, "President Obama on the Passing of Steve Jobs: 'He Changed the Way Each of Us Sees the World,'" White House archive, October 5, 2011, accessed September 10, 2018, https://obamawhitehouse.archives.gov /blog/2011/10/05/president-obama-passing-steve-jobs-he-changed-way-each-us -sees-world.

10 **The iPhone 4S:** "iPhone 4S First Weekend Sales Top Four Million," Apple, October 17, 2011, accessed September 10, 2018, www.apple.com/uk/newsroom/2011/10/17iPhone -4S-First-Weekend-Sales-Top-Four-Million.

10 **Preorders of Walter:** Andy Lewis, "Steve Jobs' Biography Sales Jump 42,000 Percent upon Death," *Hollywood Reporter,* December 5, 2011, accessed September 10, 2018, www.hollywoodreporter.com/news/steve-jobs-death-apple-biography-amazon -244747.

12 **"Come on, replace Steve?":** Lashinsky, "Apple's Tim Cook: The Genius Behind Steve Jobs."

13 **"management, even his vision":** Ty Fujimura, "Why Apple Is Doomed," *Huffington Post,* May 31, 2011, accessed September 10, 2018, www.huffingtonpost.com/ty-fujimura /why-apple-is-doomed_b_866579.html.

13 **"When Steve Jobs departed":** George Colony, "Apple = Sony," Forrester, August 4, 2017, accessed September 10, 2018, https://go.forrester.com/blogs/12-04-25-apple _sony.

14 **"The question of whether":** Adam Lashinsky, "Apple's Tim Cook Leads Different," *Fortune,* March 26, 2015, accessed September 10, 2018, http://fortune.com/2015/03/26 /tim-cook/.

14 **So widespread was:** Yukari Iwatani Kane, *Haunted Empire: Apple After Steve Jobs* (London: William Collins, 2014).

14 **"Even as he took":** Kane, *Haunted Empire,* 348.

14 **"companies, as they grow":** David Sheff, "Playboy Interview: Steve Jobs," *Playboy,* February 1985, available at Atavist, http://reprints.longform.org/playboy-interview -steve-jobs, accessed September 11, 2018.

15 **"The world was nervous":** Author interview with Greg Joswiak, March 2018.

15 **"He knew, when":** "Charlie Rose: KQED: September 13, 2014," Internet Archive, accessed September 11, 2018, https://archive.org/details/KQED_20140913_070000 _Charlie_Rose.

Chapter 2: A Worldview Shaped by the Deep South

17 **He was the second:** Michael Finch II, "Tim Cook—Apple CEO and Robertsdale's Favorite Son—Still Finds Time to Return to His Baldwin County Roots," AL.com, February 24, 2014, accessed September 11, 2018, http://blog.al.com/live/2014/02 /tim_cook_--_apple_ceo_and_robe.html.

17 **"He calls every Sunday"**: Don Cook, interview by Debbie Williams, WKRG, January 16, 2009.

18 **Don and Geraldine chose:** Finch, "Tim Cook."

18 **an area of only five square miles:** Finch, "Tim Cook."

18 **"just a little hole":** Finch, "Tim Cook."

18 **The town has had:** "Meet Our Mayor," City of Robertsdale, accessed September 11, 2018, www.robertsdale.org/mayors-office.

18 **"As a child":** "Tim Cook: Pro-discrimination 'Religious Freedom' Laws Are Dangerous," *Washington Post,* March 29, 2015, accessed September 11, 2018, www.washington post.com/opinions/pro-discrimination-religious-freedom-laws-are-dangerous-to-america/2015/03/29/bdb4ce9e-d66d-11e4-ba28-f2a685dc7f89_story.html.

19 **"I consider being gay":** Tim Cook, "Tim Cook Speaks Up," Bloomberg, October 30, 2014, accessed September 11, 2018, www.bloomberg.com/news/articles/2014-10-30/tim-cook-speaks-up.

19 **In all six years:** "Behold Tim Cook's Glory Days as 'Most Studious' in High School [Gallery]," Cult of Mac, July 27, 2015, accessed September 11, 2018, www.cultof mac.com/221717/behold-tim-cooks-glory-days-as-most-studious-in-high-school-gallery/.

19 **"He was a reliable kid":** Finch, "Tim Cook."

19 **"You didn't go around":** Finch, "Tim Cook."

19 **"He wasn't one-dimensional":** Finch, "Tim Cook."

20 **"He was just really smart":** Author interview with Clarissa Bradstock, February 2018.

20 **He played the trombone:** Finch, "Tim Cook."

21 **"the kind of person you need":** Finch, "Tim Cook."

22 **Many white families:** Yukari Iwatani Kane, *Haunted Empire: Apple After Steve Jobs* (London: William Collins, 2015), 94.

22 **Although Alabama's public:** "Robertsdale, Alabama," Wikipedia, accessed September 11, 2018, https://en.wikipedia.org/wiki/Robertsdale,_Alabama.

22 **"We had very few":** Author interview with Clarissa Bradstock, February 2018.

22 **A Baldwin County resident:** Interview by author, February 2018.

23 **While Ku Klux Klan membership:** "Ideologies," Southern Poverty Law Center, accessed September 11, 2018, www.splcenter.org/fighting-hate/extremist-files/ideology/ku-klux-klan

23 **He quickly warned:** Todd C. Frankel, "The Roots of Tim Cook's Activism Lie in Rural Alabama," *Washington Post,* March 7, 2016, accessed September 11, 2018, www.washingtonpost.com/news/the-switch/wp/2016/03/07/in-rural-alabama-the-activist-roots-of-apples-tim-cook.

23 **In a lengthy thread:** "Apple's CEO Tim Cook: An Alabama Day That Forever Changed His Life," on "Robertsdale, Past and Present" Facebook page, June 15, 2014, www.facebook.com/groups/263546476993149/permalink/863822150298909/.

24 **"They've been distributing":** Author interview with Patricia Todd, February 2018.

24 **The topic was:** Kane, *Haunted Empire,* 96.

25 **"Meeting my governor":** "GW Commencement 2015 Tim Cook," Vimeo, May 29, 2018, accessed September 12, 2018, https://vimeo.com/128073364.

25 **"I was born and raised":** "Apple CEO and Fuqua Alum Tim Cook Talks Leadership at Duke," YouTube, posted by Duke University Fuqua School of Business, accessed September 13, 2018, www.youtube.com/playlist?list=PLwEToxwSycW1uqGG -iYZOERU0WBTKIAMt.

26 **"It's just part":** Author interview with Lisa Jackson, March 2018.

26 **In a 2015 commencement:** "GW Commencement 2015 Tim Cook."

26 **"When we work":** Bryan Chaffin, "Tim Cook Soundly Rejects Politics of the NCPPR, Suggests Group Sell Apple's Stock," The Mac Observer, accessed September 13, 2018, www.macobserver.com/tmo/article/tim-cook-soundly-rejects-politics-of-the-ncppr -suggests-group-sell-apples-s.

26 **"After today's meeting":** "Tim Cook to Apple Investors: Drop Dead," National Center, November 2, 2017, accessed September 13, 2018, https://nationalcenter.org/ncppr /2014/02/28/tim-cook-to-apple-investors-drop-dead.

27 **This moral compass:** Andrew Ross Sorkin, "The Mystery of Steve Jobs's Public Giving," *New York Times,* August 29, 2011, accessed September 13, 2018, https://deal book.nytimes.com/2011/08/29/the-mystery-of-steve-jobss-public-giving/.

27 **"force for good":** "Tim Cook Wants Apple to Be a 'Force for Good,'" Cult of Mac, July 26, 2015, accessed September 13, 2018, www.cultofmac.com/251795/tim-cook -wants-apple-to-be-a-force-for-good.

27 **Its depiction of good-hearted:** Frankel, "The Roots of Tim Cook's Activism Lie in Rural Alabama."

28 **"That experience of growing up":** "Tim Cook Tells Stephen Colbert Why He Came Out as Gay," CNNMoney, accessed September 13, 2018, https://money.cnn.com /video/technology/2015/09/16/tim-cook-apple-stephen-colbert-late-show.cnnmoney /index.html.

28 **"I have to believe":** Frankel, "The Roots of Tim Cook's Activism Lie in Rural Alabama."

28 **His friend Clarissa Bradstock:** Author interview with Clarissa Bradstock, February 2018.

28 **"Robertsdale is not exactly":** Author interview with Patricia Todd, February 2018.

29 **"Chances are that":** Author interview with Baldwin County resident, February 2018.

29 **"I do think that makes":** Author interview with Patricia Todd, February 2018.

29 **"If hearing that":** "Tim Cook Speaks Up."

30 **One current resident said:** Author interview with Robertsdale resident, June 2018.

30 **"Robertsdale, Past and Present" Facebook page:** www.facebook.com/groups /263546476993149/permalink/1948196945194752/.

30 **"The jobs, that is":** Author interview with Dillan Gosnay, February 2018.

30 **Currently, Alabama doesn't:** "Employment Discrimination in Alabama," Findlaw, accessed September 13, 2018, https://corporate.findlaw.com/litigation-disputes/employment-discrimination-in-alabama.html.

30 **"Citizens of Alabama":** Associated Press, "Apple CEO Tim Cook Funds Gay Rights Initiative in Alabama," *Mercury News*, August 12, 2016, accessed September 13, 2018, www.mercurynews.com/2014/12/18/apple-ceo-tim-cook-funds-gay-rights-initiative-in-alabama.

31 **"Tim was honored":** Author interview with Patricia Todd, February 2018.

31 **In December 2014:** Associated Press, "Tim Cook Makes Personal Donation to Gay Rights Campaign," *Guardian*, December 18, 2014, accessed September 13, 2018, www.theguardian.com/technology/2014/dec/18/apple-ceo-tim-cook-donation-gay-rights-campaign.

31 **HRC's "Project One America":** "Project One America," Human Rights Campaign, accessed September 13, 2018, www.hrc.org/campaigns/project-one-america.

31 **Since Cook's donation, HRC:** "About Us," Human Rights Campaign, accessed September 13, 2018, www.hrc.org/hrc-story/about-us.

31 **"He has contributed":** Author interview with Patricia Todd, February 2018.

31 **"If you'd told me":** Author interview with Patricia Todd, February 2018.

32 **"Most of my formative":** Erin Edgemon, "Apple's Tim Cook Talks MLK, Auburn, Coding During Birmingham Visit," AL.com, April 4, 2018, accessed September 13, 2018, www.al.com/news/birmingham/index.ssf/2018/04/tim_cook_talks_mlk_auburn_codi.html.

32 **"Ever since he was":** Finch, "Tim Cook."

32 **"The well-to-do people":** Hanno van der Bijl, "Apple's Tim Cook on Leadership, Workplace Diversity, Alabama and Auburn Rivalry," *Birmingham Business Journal*, June 1, 2018, accessed September 13, 2018, www.bizjournals.com/birmingham/news/2018/04/06/tim-cook-on-leadership-workplace-diversity-alabama.html.

33 **"He could cut through":** Kane, *Haunted Empire*, 98.

33 **"I don't deserve this":** Kane, *Haunted Empire*, 99.

33 **"a very quiet":** Miguel Helft, "Tim Cook Is Running Apple, but Not Imitating Steve Jobs," *New York Times*, January 23, 2011, accessed September 13, 2018, www.nytimes.com/2011/01/24/technology/24cook.html.

33 **Now he jokes that:** Jasper Hamill, "Apple CEO Tim Cook Reveals How YOU Can Follow in His Footsteps," *The Sun*, October 13, 2017, accessed September 13, 2018, www.thesun.co.uk/tech/4663185/apple-ceo-tim-cook-reveals-a-big-career-secret-and-tells-how-you-can-follow-in-his-footsteps.

34 **"I believe this is":** "The Auburn Creed," Auburn University, accessed September 13, 2018, www.auburn.edu/main/welcome/creed.html.

34 **"Though the sentiment":** "Auburn University Spring 2010 Commencement Speaker Tim Cook," YouTube, posted by Auburn University, May 18, 2010, accessed September 13, 2018, www.youtube.com/watch?v=xEAXuHvzjao.

34 **He enrolled in a cooperative:** Kane, *Haunted Empire*, 99.

35 **"The truth is":** Kane, *Haunted Empire*, 99.

Chapter 3: Learning the Trade at Big Blue

37 **The home computer industry:** U.S. Census Bureau, *Home Computers and Internet Use in the United States: August 2000* (Washington, DC: U.S. Department of Commerce, September 2001), www.census.gov/prod/2001pubs/p23-207.pdf.

37 **more than 350,000 people:** "1981," IBM—Archives—History of IBM—United States, accessed September 13, 2018, www-03.ibm.com/ibm/history/history/year _1981.html.

38 **The $1,565 Personal Computer:** "IBM Personal Computer," Wikipedia, accessed September 13, 2018, https://en.wikipedia.org/wiki/IBM_Personal_Computer.

38 **Take, for example:** "Apple Watch Series 3—Technical Specifications," Apple, accessed September 13, 2018, https://support.apple.com/kb/sp766?locale=en_US.

38 **One of the first brochures:** "Full Text of 'Brochure: IBM Personal Computer (PC),'" accessed September 13, 2018, https://archive.org/stream/1982-ibm-personal-computer /1982-ibm-personal-computer_djvu.txt.

39 **By the end of that year:** "The Birth of the IBM PC," IBM—Archives—History of IBM—United States, accessed September 13, 2018, www-03.ibm.com/ibm/history /exhibits/pc25/pc25_birth.html.

39 **The company had originally estimated:** "Encyclopedia," *PC* magazine, accessed September 13, 2018, www.pcmag.com/encyclopedia/term/44650/ibm-pc.

39 **In 1982, IBM:** "Personal Computer Market Share: 1975–2004," accessed September 13, 2018, www.retrocomputing.net/info/siti/total_share.html.

39 **"There are some occasions":** Otto Friedrich, "The Computer Moves In," *Time*, January 3, 1983, accessed September 13, 2018, http://content.time.com/time/subscriber /article/0,33009,953632-3,00.html.

39 **"They FedExed me the magazine":** Walter Isaacson, *Steve Jobs* (New York: Simon & Schuster, 2011), 567.

40 **When he joined IBM:** Author interview with Dick Daugherty, February 2018.

40 **Every single day:** Author interview with Gene Addesso, February 2018.

41 **When the process was complete:** Author interview with Gene Addesso, February 2018.

41 **"Just-in-time manufacturing meant":** Author interview with Gene Addesso, February 2018.

41 **The JIT philosophy:** "Just-in-Time Manufacturing," Wikipedia, accessed September 13, 2018, https://en.wikipedia.org/wiki/Just-in-time_manufacturing.

41 **"a flow process":** Taiichi Ohno, *Toyota Production System: Beyond Large-Scale Production* (London: CRC Press, 2014), 4.

42 **"Sometimes, of course":** Ohno, *Toyota Production System*, 26.

42 **"We have found in":** Henry Ford, *My Life and Work: An Autobiography of Henry*

Ford (Greenbook Publications, 2010), Kindle edition, https://www.amazon.com/dp/B00306KYVQ/.

43 **Employee morale had increased:** Deby Veneziale, "Workshop Report: Continuous Flow Manufacturing at IBM Tucson," Summer 1989, www.ame.org/sites/default/files/target_articles/89Q2A3.pdf.

43 **"It was very difficult":** Author interview with Dick Daugherty, February 2018.

44 **"I had him ranked":** Author interview with Dick Daugherty, February 2018.

44 **"The thing that is impressive":** Author interview with Ray Mays, March 2018.

44 **"One of the things we do":** Author interview with Ray Mays, March 2018.

45 **"guess who was chosen":** Author interview with Dick Daugherty, February 2018.

45 **"You could look at this guy":** Author interview with Gene Addesso, February 2018.

45 **"It was a public relations operation":** Author interview with Ray Mays, March 2018.

46 **"You work all day":** Author interview with Ray Mays, March 2018.

46 **"He's a good business mind":** Author interview with Greg Joswiak, March 2018.

47 **"When many people in business":** "Apple CEO and Fuqua Alum Tim Cook Talks Leadership at Duke," YouTube, posted by Duke University Fuqua School of Business, accessed September 13, 2018, www.youtube.com/playlist?list=PLwEToxwSycW1uqGG-iYZOERU0WBTKIAMt.

48 **"a group of people":** Author interview with Gene Addesso, February 2018.

48 **"He just was very thoughtful":** Author interview with Dick Daugherty, February 2018.

48 **"people . . . enjoy[ed]":** Author interview with Ray Mays, March 2018.

48 **Addesso said he kept:** Author interview with Gene Addesso, February 2018.

48 **"I never knew":** Author interview with Gene Addesso, February 2018.

49 **"We were scrambling":** Author interview with Dick Daugherty, February 2018.

49 **"We had every available":** Author interview with Ray Mays, March 2018.

50 **The company is long gone:** Edward O. Welles, "When a Billion-Dollar Company Ain't Enough," *Inc.*, May 1, 1995, accessed September 13, 2018, www.inc.com/magazine/19950501/2265.html.

50 **In his first year:** Edgar Online, Intelligent Electronics Inc Form Def 14A, July 23, 1996, http://b4utrade.brand.edgar-online.com/efxapi/EFX_dll/EDGARpro.dll?FetchFilingCONVPDF1?SessionID=GReEUbNUi2_slft&ID=1469265.

50 **"The move was so lucrative":** Author interview with Ray Mays, March 2018.

50 **The news made him:** Yukari Iwatani Kane, "The Job After Steve Jobs: Tim Cook and Apple," *Wall Street Journal*, March 1, 2014, accessed September 13, 2018, www.wsj.com/articles/the-job-after-steve-jobs-tim-cook-and-apple-1393637952.

51 **It turns out that he:** Kane, "The Job After Steve Jobs."

51 **In February 1997, Compaq:** "Compaq Offers Cheap PCs," CNNMoney, accessed September 13, 2018, https://money.cnn.com/1997/02/20/technology/compaq/.

52 **The Cyrix chip was from:** "The Secret History of the Sub-$1,000 Computer," CNET,

accessed October 4, 2018, https://www.cnet.com/news/the-secret-history-of-the-sub
-1000-computer/

52 **The Presario 2000 lineup:** "Compaq 4Q Net Grows," CNNMoney, accessed September 13, 2018, https://money.cnn.com/1999/01/27/companies/compaq/.

52 **"We are sending a shockwave":** "Compaq Launches New Business Model Creating Customer Value Revolution," Business Wire, July 10, 1997.

53 **"By continually finetuning our Optimized Distribution Model":** *Business Wire* press release in Nexis, "Compaq Lowers Prices Across It Entire Deskpro Line by Up to 18 Percent," February 2, 1998.

53 **"From Compaq's perspective":** Peter C. Y. Chow and Gill Bates, eds., *Weathering the Storm: Taiwan, Its Neighbors, and the Asian Financial Crisis* (Washington, DC: Brookings Institution, 2000), 181.

53 **"Tim Cook came out of procurement":** Isaacson, *Steve Jobs*, 360.

Chapter 4: A Once-in-a-Lifetime Opportunity to Join a Near-Bankrupt Company

55 **When Cook joined Apple:** "Tim Cook Joins Apple as Senior Vice President of Worldwide Operations," March 11, 1998, available at Internet Archive, accessed September 13, 2018, web.archive.org/web/19980429150102/http://www.apple.com:80/pr/library/1998/mar/11org.html.

55 **Steve Jobs had recently rejoined:** Mark Leibovich, "Jobs Drops 'Interim' Title, Apple Chief Executive Affirms His Commitment," *Washington Post*, January 6, 2000, accessed September 13, 2018, www.washingtonpost.com/wp-srv/WPcap/2000-01/06/014r-010600-idx.html.

56 **The company employed:** Apple Inc., "SEC Filings," Form 10-K, December 1994, accessed September 13, 2018, http://investor.apple.com/secfiling.cfm?filingid=320193-94-16&cik=320193.

56 **Windows 95 was a huge commercial success:** Jonathan Chew, "Microsoft Launched This Product 20 Years Ago and Changed the World," *Fortune*, August 24, 2015, accessed September 13, 2018, http://fortune.com/2015/08/24/20-years-microsoft-windows-95/.

56 **Apple had made a profit:** Reuters, "iMac, Therefore I Make Money," *Wired*, October 12, 1998, accessed September 13, 2018, www.wired.com/1998/10/imac-therefore-i-make-money/.

56 **it reported a loss of $69 million:** "Apple Loss Hits $69 Million," CNET, January 18, 1996, accessed September 13, 2018, www.cnet.com/news/apple-loss-hits-69-million/.

56 **This was followed by a much bigger loss:** Jim Carlton, "Apple Sees $700 Million Loss in Quarter After Write-Downs," *Wall Street Journal*, March 29, 1996, accessed September 13, 2018, www.wsj.com/articles/SB868490956869493000.

56 **As a result, Apple started laying off:** Elizabeth Corcoran, "Spindler Is Out at

Apple," *Washington Post*, February 3, 1996, accessed September 13, 2018, www.washingtonpost.com/archive/business/1996/02/03/spindler-is-out-at-apple/4ec75ebd-8d56-43aa-8a7c-4320314ca30b/?utm_term=.cb58e4aea0c4.

56 **In the eighteen months Amelio:** Steve Lohr, "Apple Computer Ousts Chief in Response to Poor Results," *New York Times*, July 10, 1997, accessed September 13, 2018, www.nytimes.com/1997/07/10/business/apple-computer-ousts-chief-in-response-to-poor-results.html.

56 **But Amelio did one thing right:** John Markoff, "Steven Jobs Making Move Back to Apple," *New York Times*, December 21, 1996, accessed September 13, 2018, www.nytimes.com/1996/12/21/business/steven-jobs-making-move-back-to-apple.html.

57 **"If I couldn't figure this out":** "Steve Jobs WWDC 1998 Keynote (Part 1)," YouTube, posted by AppleKeynotes, October 23, 2007, accessed September 13, 2018, www.youtube.com/watch?v=YJGcJgpOU9w.

57 **One board member said the plan:** P. Burrows and R. Grover, "Steve Jobs' Magic Kingdom," *Bloomberg Businessweek*, February 6, 2006, accessed September 13, 2018, www.bloomberg.com/news/articles/2006-02-05/steve-jobs-magic-kingdom.

57 **His neighbors in Palo Alto:** Alan Deutschman, *The Second Coming of Steve Jobs* (Milsons Point, NSW: Random House International, 2001), 261.

58 **The mischievous employee:** Yukari Iwatani Kane, *Haunted Empire: Apple After Steve Jobs* (London: William Collins, 2015).

58 **Grueling work hours:** Deutschman, *The Second Coming of Steve Jobs*, 257.

59 **In 1993, it was burned:** James Daly, "Apple Excess Inventory Spawns Macintosh Auctions," *ComputerWorld*, November 15, 1993, available at Google Books, https://books.google.com/books?id=Od_7AEHBZvgC&pg=PA61&lpg=PA61#v=onepage&q&f=false.

59 **Forecasts were too low:** Mike Langberg, "1995: Is Apple Walking the Wrong Path?," *Mercury News*, October 2, 1995, accessed September 13, 2018, www.mercurynews.com/2014/08/29/1995-is-apple-walking-the-wrong-path.

59 **Preorders for the new:** Tom Quinlan, "Power Macs an Instant Hit with Apple's Core Following," *Infoworld*, March 21, 1994, available at Google Books, https://books.google.ca/books?id=DTsEAAAAMBAJ&lpg=PA22&pg=PA33#v=onepage&q&f=false.

59 **"Not only has Apple finally regained":** "Power Macintosh 6100/60" (review), *Macworld*, no. 9406, June 1994, available at https://archive.org/stream/MacWorld_9406_June_1994#page/n57/mode/2up, accessed September 13, 2018.

59 **The *San Francisco Chronicle* called:** "Power Mac Shortage Bruises Apple," *San Francisco Chronicle*, March 22, 1995, accessed September 13, 2018, www.sfgate.com/business/article/Power-Mac-Shortage-Bruises-Apple-3040630.php.

59 **"[Apple] was unable":** Robert B. Handfield and Ernest L. Nichols Jr., *Supply Chain Redesign: Transforming Supply Chains into Integrated Value Systems* (Upper Saddle River, NJ: FT Press, 2002), 129.

59 **It was one of Apple's worst years:** David Kiger, "Apple and the 1995 Disaster: What

Happened and How They Recovered," *David Kiger* (blog), July 18, 2016, accessed September 13, 2018, http://aboutdavidkiger.net/apple-1995-disaster-happened -recovered/.

60 **The industry publication *Supply Chain Digest*:** "The 11 Greatest Supply Chain Disasters," *Supply· Chain Digest,* January 2006, www.scdigest.com/assets/reps /SCDigest_Top-11-SupplyChainDisasters.pdf.

60 **Today Apple is investing:** Nate Lanxon, "Apple Supplier Dialog Falls on Report of In-House Chip Move," Bloomberg, November 30, 2017, accessed September 13, 2018, www.bloomberg.com/news/articles/2017-11-30/apple-reportedly-making-in-house -power-chips-in-blow-to-dialog.

60 **"we were a little timid":** Bloomberg News, "Company News; Apple Discloses Shortages of High-End Units," *New York Times,* March 22, 1995, accessed September 13, 2018, www.nytimes.com/1995/03/22/business/company-news-apple -discloses-shortages-of-high-end-units.html.

60 **"Investors just hate it":** "Power Mac Shortage Bruises Apple," *San Francisco Chronicle,* March 22, 1995, accessed September 13, 2018, www.sfgate.com/business/article /Power-Mac-Shortage-Bruises-Apple-3040630.php.

61 **First, in 1996:** Joel West, *Apple Computer: The iCEO Seizes the Internet* (Irvine, CA: Personal Computing Industry Center, October 2002), ftp://ftp.apple.asimov.net /pub/apple_II/documentation/misc/APPLE_Computer_Inc._Intro_Article_3.pdf.

61 **"Apple sold a circuit board factory":** West, *Apple Computer: The iCEO Seizes the Internet.*

61 **"He became a manager":** Isaacson, *Steve Jobs* (New York: Simon & Schuster, 2011), 573.

62 **Jobs denigrated Dell:** "Today in Apple History: Michael Dell Says He'd Shut Down Apple," Cult of Mac, https://www.cultofmac.com/448147/today-apple-history-michael -dell-says-hed-shut-apple-refund-shareholders/.

62 **"Steve created the whole industry":** Sam Colt, "Tim Cook Gave His Most In-Depth Interview to Date—Here's What He Said," Business Insider, September 20, 2014, accessed September 13, 2018, www.businessinsider.com/tim-cook-full-interview -with-charlie-rose-with-transcript-2014-9.

63 **Cook came away from the meeting:** "Tim Cook on Apple TV (Sept 12, 2014) | Charlie Rose Show," YouTube, posted by Charlie Rose, September 12, 2014, accessed September 13, 2018, www.youtube.com/watch?v=oBMo8Oz9jsQ.

63 **"One CEO I consulted":** "Auburn University Spring 2010 Commencement Speaker Tim Cook," YouTube, posted by Auburn University, May 18, 2010, accessed September 13, 2018, www.youtube.com/watch?v=xEAXuHvzjao.

63 **"there was literally no one":** Colt, "Tim Cook Gave His Most In-Depth Interview to Date."

63 **"I'd always thought that":** Colt, "Tim Cook Gave His Most In-Depth Interview to Date."

64 **"Five minutes into my initial interview":** Isaacson, *Steve Jobs,* 573.

64 **"Working at Apple":** "Auburn University Spring 2010 Commencement Speaker Tim Cook."

64 **Jobs had found a partner:** Isaacson, *Steve Jobs*, 573.

64 **Jobs hired Cook:** Doug Bartholomew, "What's Really Driving Apple's Recovery," *IndustryWeek*, March 16, 1999, accessed September 13, 2018, www.industryweek .com/companies-amp-executives/whats-really-driving-apples-recovery.

64 **"I remember when Steve":** Author interview with Greg Joswiak, March 2018.

65 **"As you can imagine":** West, *Apple Computer: The iCEO Seizes the Internet.*

65 **"You have to have that business sense":** Author interview with Greg Joswiak, March 2018.

65 **"To this day I remember":** Author interview with Deirdre O'Brien, March 2018.

65 **She is now Apple's head:** Author interview with Deirdre O'Brien, March 2018.

67 **Just seven months after arriving:** Owen Thomas, "Apple's Recipe: Just One Cook," *Owen Thomas Is Still in Beta* (blog archive), March 21, 2018, accessed September 13, 2018, https://owenthomas.wordpress.com/1998/10/16/apples-recipe-just-one -cook/.

68 **Now Quanta Computer:** West, *Apple Computer: The iCEO Seizes the Internet.*

68 **Big stockpiles of unsold computers:** David Bovet and Joseph Martha, "Value Nets: Reinventing the Rusty Supply Chain for Competitive Advantage," *Strategy & Leadership* 28, no. 4 (July 1, 2000): 21–26. doi:10.1108/10878570010378654.

68 **"Warehouses tend to collect":** Bovet and Martha, "Value Nets."

68 **"You want to manage it":** Adam Lashinsky, "Apple's Tim Cook: The Genius Behind Steve Jobs," *Fortune*, November 24, 2008, accessed September 10, 2018, http://fortune .com/2008/11/24/apple-the-genius-behind-steve/.

69 **Parts were ordered:** Bartholomew, "What's Really Driving Apple's Recovery."

69 **In the seven months:** Thomas, "Apple's Recipe: Just One Cook."

69 **In 1998, Cook got rid:** Brent Schlender and Rick Tetzeli, *Becoming Steve Jobs: The Evolution of a Reckless Upstart into a Visionary Leader* (Toronto: Signal, 2016), 223.

70 **To make sure:** Paul Simpson, "Tim Cook: The 'Cool Customer' Behind Apple's Supply Chain Success," *Supply Management*, January 18, 2016, accessed September 13, 2018, www.cips.org/supply-management/analysis/2016/february/tim-cook-the-cool -customer-behind-apples-supply-chain-success/.

70 **Not only did Apple:** Simpson, "Tim Cook: The 'Cool Customer' Behind Apple's Supply Chain Success."

71 **"all the designers":** Author interview with Gautam Baksi, April 2013.

Chapter 5: Saving Apple Through Outsourcing

73 **After reporting a net loss:** "Apple 4Q Caps Profitable 1998," CNNMoney, accessed September 13, 2018, https://money.cnn.com/1998/10/14/technology/apple/.

73 **"Apple grew faster":** "Apple 4Q Caps Profitable 1998."

74 **"worked very hard":** Author interview with Deirdre O'Brien, March 2018.

74 **"It was not a traditional":** Author interview with Deirdre O'Brien, March 2018.

75 **"This is state of the art":** Leander Kahney, "Ex-Apple Engineer Tells How the Company's Manufacturing Works," Cult of Mac, June 29, 2017, accessed September 13, 2018, www.cultofmac.com/488624/revolutionizing-manufacturing-using-machine-learning-podcast-transcript/.

75 **At Foxconn and other assembly plants:** Kahney, "Ex-Apple Engineer Tells How the Company's Manufacturing Works."

75 **They are more like factory towns:** John Vause, "Inside China Factory Hit by Suicides," CNN, June 2, 2010, accessed September 13, 2018, http://edition.cnn.com/2010/WORLD/asiapcf/06/01/china.foxconn.inside.factory/index.html.

76 **It also has plants:** Author interview with Duane O'Very, April 2018.

76 **Foxconn demonstrated this:** Charles Duhigg and Keith Bradsher, "How the U.S. Lost Out on iPhone Work," *New York Times*, January 21, 2012, accessed September 13, 2018, www.nytimes.com/2012/01/22/business/apple-america-and-a-squeezed-middle-class.html.

76 **More than eight thousand workers:** Duhigg and Bradsher, "How the U.S. Lost Out on iPhone Work."

77 **"That's very easy":** Author interview with Gautum Baksi, April 2013.

77 **"They could hire 3,000":** Duhigg and Bradsher, "How the U.S. Lost Out on iPhone Work."

78 **"Now it seems very clear":** Author interview with Deirdre O'Brien, March 2018.

79 **In September 1999:** Alan Deutschman, *The Second Coming of Steve Jobs* (Milsons Point, NSW: Random House International, 2001), 279.

79 **In 2004, Jobs appointed him:** "Tim Cook Named COO of Apple," Apple, October 14, 2005, accessed September 13, 2018, www.apple.com/newsroom/2005/10/14Tim-Cook-Named-COO-of-Apple/.

79 **"Tim has been doing this":** "Tim Cook Named COO of Apple."

80 **Steve Jobs was the kind:** Walter Isaacson, *Steve Jobs* (New York: Simon & Schuster, 2011), 359.

81 **"He's a very quiet leader":** Author interview with Greg Joswiak at Apple Park, March 2018.

81 **"He'll ask you ten questions":** Adam Lashinsky, "Apple's Tim Cook: The Genius Behind Steve Jobs," *Fortune*, November 24, 2008, accessed September 10, 2018, http://fortune.com/2008/11/24/apple-the-genius-behind-steve/.

81 **"They're nervous going into":** Adam Lashinsky, *Inside Apple: How America's Most Admired—and Secretive—Company Really Works* (London: John Murray, 2012), 220.

81 **"That number is wrong":** Yukari Iwatani Kane, *Haunted Empire: Apple After Steve Jobs* (London: William Collins, 2015), 101.

82 **"They want to know":** Author interview with Helen Wang, March 2018.

82 **"everything is possible":** Author interview with Helen Wang, March 2018.

82 **"The senior director level":** Author interview with Helen Wang, March 2018.

83 **He didn't even stop:** Lashinsky, "Apple's Tim Cook: The Genius Behind Steve Jobs."

83 **"I would get a couple":** Author interview with Bruce Sewell, March 2018.

84 **"He exercises a lot":** Author interview with Bruce Sewell, March 2018.

84 **Work was (and remains):** Lashinsky, "Apple's Tim Cook: The Genius Behind Steve Jobs."

84 **"I don't like to lose":** Kane, *Haunted Empire*, 103.

84 **"In business, as in sports":** "Auburn University Spring 2010 Commencement Speaker Tim Cook," YouTube, posted by Auburn University, May 18, 2010, accessed September 13, 2018, www.youtube.com/watch?v=xEAXuHvzjao.

84 **"I know he and the rest":** "Letter from Steve Jobs to Apple Employees," Reuters, January 15, 2009, accessed September 13, 2018, www.reuters.com/article/us-steve-jobs-letter-sb-idUSTRE50D7JG20090115.

84 **During Jobs's absence:** "Apple Sells One Million iPhone 3Gs in First Weekend," Apple, July 14, 2008, accessed September 13, 2018, www.apple.com/newsroom/2008/07/14Apple-Sells-One-Million-iPhone-3Gs-in-First-Weekend/.

85 **Jobs returned to Apple:** "Steve Jobs: 'I'm Vertical,'" *Entertainment Weekly*, September 9, 2009, accessed September 13, 2018, https://ew.com/article/2009/09/09/steve-jobs-im-vertical/.

85 **Cook did such a good job:** David Goldman, "Steve Jobs Takes Medical Leave of Absence," CNNMoney, January 17, 2011, accessed September 13, 2018, https://money.cnn.com/2011/01/17/technology/steve_jobs_leave/index.htm.

Chapter 6: Stepping into Steve Jobs's Shoes

87 **Headlining his first Apple keynote:** "Apple Special Event March 7 2012 iPad 3 the New iPad Full Apple Keynote March 2012 (Full)," YouTube, posted by kent880821, March 8, 2012, accessed September 13, 2018, www.youtube.com/watch?v=z5yCqaf9yBc.

89 **It wasn't a big deal:** Matthew Panzarino, "Apple 'Misses' in Q3, but Beats Own Estimates with Record iPad Sales," The Next Web, July 25, 2012, accessed September 13, 2018, https://thenextweb.com/apple/2012/07/24/apple-q3-2012-35b-revenue-8-8b-profit-with-9-32-eps-26m-iphones-17m-ipads-sold/.

89 **In May 2012:** Millward Brown, "Top Brands Thrive Despite Economy According to Millward Brown's Latest BrandZ[TM] Top 100 Most Valuable Global Brands Study," PR Newswire, May 2, 2012, accessed September 13, 2018, www.prnewswire.com/news-releases/top-brands-thrive-despite-economy-according-to-millward-browns-latest-brandztm-top-100-most-valuable-global-brands-study-152379145.html.

89 **"Now worth more than":** Jon Russell, "Apple Ranked Top as Tech Firms Dominate Global Brand Report," The Next Web, May 22, 2012, accessed September 13, 2018,

https://thenextweb.com/apple/2012/05/22/apple-beats-ibm-and-google-to-top
-global-brand-report-as-tech-firms-dominate/.

89 **Oh had been in the job:** *Korea Herald*/Asia News Network, "Apple Korea Chief Gets
Pink Slip by Email," accessed September 13, 2018, http://www.koreaherald.com/view
.php?ud=20121024000338.

89 **In July 2012:** Philip Elmer-Dewitt, "Apple's MobileMe Is Dead—but You Can Still
Retrieve Your Files," *Fortune*, July 1, 2012, accessed September 13, 2018, http://
fortune.com/2012/07/01/apples-mobileme-is-dead-but-you-can-still-retrieve
-your-files/.

89 **At the end of September:** Christina Bonnington, "So Long, Ping: Apple Shuttering
Failed Social Network Sept. 30," *Wired*, September 13, 2012, accessed September 13,
2018, www.wired.com/2012/09/goodbye-ping/.

90 **didn't "need to have":** Macworld Staff, "Tim Cook at D10: In His Own Words,"
Macworld, May 29, 2012, accessed September 13, 2018, www.macworld.com/article
/1167008/tim_cook_at_d10_in_his_own_words.html?page=2.

90 **The year 2012 also saw:** "Apple Announces Changes to Increase Collaboration
Across Hardware, Software & Services," Apple, October 29, 2012, accessed September
13, 2018, www.apple.com/newsroom/2012/10/29Apple-Announces-Changes-to
-Increase-Collaboration-Across-Hardware-Software-Services/.

90 **"I don't do lunch":** Rupert Neate, "Dixons Boss John Browett Swaps Hemel Hemp-
stead for Apple's California HQ," *Guardian*, January 31, 2012, accessed September
13, 2018, www.theguardian.com/business/2012/jan/31/dixons-boss-john-browett
-apple.

91 **"I talked to many people":** Kevin Rawlinson, "Former Dixons Executive John
Browett Shown the Door at Apple," *Independent*, October 31, 2012, accessed September 13, 2018, www.independent.co.uk/life-style/gadgets-and-tech/news/former
-dixons-executive-john-browett-shown-the-door-at-apple-8262120.html.

91 **"Making these changes":** Rawlinson, "Former Dixons Executive John Browett Shown
the Door at Apple."

92 **"I just didn't fit":** James Titcomb, "Why New Apple Retail Chief's British Predecessor John Browett Was Fired," *Telegraph*, October 15, 2013, accessed September 13,
2018, www.telegraph.co.uk/finance/newsbysector/retailandconsumer/10379517
/Why-new-Apple-retail-chiefs-British-predecessor-John-Browett-was-fired.html.

92 **A 2011 *Bloomberg Businessweek* profile:** Adam Satariano, Peter Burrows, and
Brad Stone, "Scott Forstall, the Sorcerer's Apprentice at Apple," *Bloomberg Businessweek*, October 13, 2011, accessed September 10, 2018, www.bloomberg.com/news
/articles/2011-10-12/scott-forstall-the-sorcerers-apprentice-at-apple.

92 ***Fortune* reporter Adam Lashinky's:** Philip Elmer-Dewitt, "Scott Forstall Is Apple's
'CEO-in-Waiting' Says New Book," *Fortune*, January 17, 2012, accessed September
10, 2018, http://fortune.com/2012/01/17/scott-forstall-is-apples-ceo-in-waiting-says
-new-book/.

93 **When Forstall introduced Siri:** "Apple Knowledge Navigator Video (1987)," YouTube, posted by Mac History, March 4, 2012, accessed September 13, 2018, www.youtube.com/watch?v=umJsITGzXd0.

93 **But when Apple launched:** Todd Wasserman, "Wozniak: Siri Was Better Before Apple Bought It," Mashable, June 15, 2012, accessed September 13, 2018, https://mashable.com/2012/06/15/wozniak-on-siri/#EepIlr_2_8q8.

93 **The pair cofounded:** Cromwell Schubarth, "Now We Know What Siri's Creators and Investors Got When They Sold Viv to Samsung," *Silicon Valley Business Journal,* March 1, 2017, accessed September 13, 2018, www.bizjournals.com/sanjose/news/2017/03/01/now-we-know-what-siris-creators-and-investors-got.html.

94 **Apple's mapping software:** "Apple Previews iOS 6 with All New Maps, Siri Features, Facebook Integration, Shared Photo Streams & New Passbook App," Apple, August 31, 2018, accessed September 13, 2018, www.apple.com/newsroom/2012/06/11Apple-Previews-iOS-6-With-All-New-Maps-Siri-Features-Facebook-Integration-Shared-Photo-Streams-New-Passbook-App/.

94 **The *New York Times*:** David Pogue, "Apple's New Maps App Is Upgraded, but Full of Snags—Review," *New York Times,* September 26, 2012, accessed September 13, 2018, www.nytimes.com/2012/09/27/technology/personaltech/apples-new-maps-app-is-upgraded-but-full-of-snags-review.html.

94 **And an Apple executive:** Yukari Iwatani Kane, *Haunted Empire: Apple After Steve Jobs* (London: William Collins, 2015), 236.

95 **An article by:** Jay Yarow, "The Apple Maps Disaster Is Really Bad News for Apple's 'CEO-in-Waiting,'" Business Insider, September 28, 2012, accessed September 13, 2018, www.businessinsider.com/scott-forstall-apple-maps-2012-9.

95 **"At Apple, we strive":** Brian X. Chen, "Tim Cook Apologizes for Apple's Maps," *New York Times,* September 28, 2012, accessed September 13, 2018, https://bits.blogs.nytimes.com/2012/09/28/tim-cook-maps/.

95 **"a form of abasement":** "Tim Cook's Apology for Apple Maps: Proof He's No Steve Jobs?," *The Week,* September 28, 2012, accessed September 13, 2018, http://theweek.com/articles/471907/tim-cooks-apology-apple-maps-proof-hes-no-steve-jobs.

96 **A couple of years prior:** Doug Gross, "Apple on iPhone Complaints: You're Holding It Wrong," CNN, June 25, 2010, accessed September 13, 2018, www.cnn.com/2010/TECH/mobile/06/25/iphone.problems.response/index.html.

96 **"bold and decisive":** Author interview with Greg Joswiak, March 2018.

96 **Internally, his departure:** Leo Kelion, "Tony Fadell: From iPod Father to Thermostat Start-up," BBC News, November 29, 2012, accessed September 13, 2018, www.businessinsider.com/ipod-inventor-fired-apple-exec-scott-forstall-got-what-he-deserved-2012-11.

96 **"He wanted to be involved":** Author interview with former Apple employee, September 2014.

97 **"So, I think Apple":** Kelion, "Tony Fadell: From iPod Father to Thermostat Start-up."

97 **"As much as he was contributing"**: Author interview with Horace Dediu, March 2018.

97 **"Tim felt that"**: Author interview with Horace Dediu, March 2018.

97 **Cook said the management:** Josh J. Tyrangiel, "Tim Cook's Freshman Year: The Apple CEO Speaks," Bloomberg, December 6, 2012, accessed September 13, 2018, www.bloomberg.com/news/articles/2012-12-06/tim-cooks-freshman-year-the -apple-ceo-speaks.

98 **"You have to be"**: Tyrangiel, "Tim Cook's Freshman Year."

98 **"We want this fast-moving"**: Tyrangiel, "Tim Cook's Freshman Year."

98 **"The decision to dump"**: Dan Crow, "We've Passed Peak Apple: It's All Downhill from Here," *Guardian,* November 7, 2012, accessed September 13, 2018, www.the guardian.com/technology/2012/nov/07/peak-apple.

99 **In January 2013:** Ian Sherr and Evan Ramstad, "Has Apple Lost Its Cool to Samsung?," *Wall Street Journal,* January 28, 2013, accessed September 14, 2018, www .wsj.com/articles/SB10001424127887323854904578264090074879024.

99 **"We understand that this"**: Jay Yarow, "Phil Schiller Exploded on Apple's Ad Agency in an Email," Business Insider, April 7, 2014, accessed September 14, 2018, www.businessinsider.com/phil-schiller-emails-2014-4.

99 **"This is not 1997"**: Yarow, "Phil Schiller Exploded on Apple's Ad Agency in an Email."

100 **Since 2006, the company:** "(PRODUCT)RED™," Apple, accessed September 14, 2018, www.apple.com/product-red/.

100 **He plans to donate:** Adam Lashinsky, "Apple's Tim Cook Leads Different," *Fortune,* March 25, 2015, http://fortune.com/2015/03/26/tim-cook.

100 **Apple had also donated:** "Apple Support for Charity," MacRumors.com Help Center, accessed September 14, 2018, https://macrumors.zendesk.com/hc/en-us/articles /202084918-Apple-Support-for-Charity.

101 **The bid raised $610,000:** "Robert F. Kennedy Center's 6th Annual Spring Auction Raises over $1.1 Million at Charitybuzz.com," Charitybuzz, May 15, 2013, accessed September 14, 2018, www.charitybuzz.com/press_releases/646.

101 **"I'm thrilled to announce"**: Dawn Chmielewski, "Apple's Holiday Product Red Campaign Raises $20 Million for AIDS Research," Recode, December 17, 2014, accessed September 14, 2018, www.recode.net/2014/12/17/11633904/apples-holiday -product-red-campaign-raises-20-million-for-aids.

102 **But in 2018:** "Apple Employees Get Big Bonus Thanks to Trump," Cult of Mac, January 19, 2018, accessed September 14, 2018, www.cultofmac.com/523889/apple -employees-get-big-bonus-thanks-trump/.

102 **"Apple's charity efforts"**: Chance Miller, "Here's the Full Email Tim Cook Sent to Apple Employees Announcing Bonuses & New Charity Donation Matching," 9to5Mac, January 17, 2018, accessed September 14, 2018, https://9to5mac.com/2018/01/17/tim -cook-bonus-email-to-employees/.

102 **In February 2012:** Bill Weir, "A Trip to the iFactory: 'Nightline' Gets an Unprecedented Glimpse Inside Apple's Chinese Core," ABC News, February 20, 2012, accessed September 14, 2018, https://abcnews.go.com/International/trip-ifactory-nightline -unprecedented-glimpse-inside-apples-chinese/story?id=15748745.

103 **The *New York Times* also published:** "The iEconomy" (article series), *New York Times,* accessed September 14, 2018, http://archive.nytimes.com/www.nytimes.com /interactive/business/ieconomy.html.

103 **"We care about every worker":** Mark Gurman, "Tim Cook Responds to Claims of Factory Worker Mistreatment: 'We Care About Every Worker in Our Supply Chain,'" 9to5Mac, January 27, 2012, accessed September 14, 2018, https://9to5mac.com /2012/01/26/tim-cook-responds-to-claims-of-factory-worker-mistreatment-we-care -about-every-worker-in-our-supply-chain/.

104 **"Apple takes working conditions":** Philip Elmer-Dewitt, "Transcript: Apple CEO Tim Cook at Goldman Sachs," *Fortune,* February 15, 2012, accessed September 14, 2018, http://fortune.com/2012/02/15/transcript-apple-ceo-tim-cook-at-goldman -sachs/.

104 **Cook's pledge of meaningful reform:** Nick Wingfield, "Apple's Chief Puts Stamp on Labor Issues," *New York Times,* April 2, 2012, accessed September 14, 2018, www .nytimes.com/2012/04/02/technology/apple-presses-its-suppliers-to-improve -conditions.html.

104 **"It looks like a pattern":** Wingfield, "Apple's Chief Puts Stamp on Labor Issues."

105 **Despite Cook's pledges:** Maxim Duncan, "Foxconn China Plant Closed After 2,000 Riot," Reuters, September 24, 2012, accessed September 14, 2018, www.reuters.com /article/us-hon-hai/foxconn-china-plant-closed-after-2000-riot-idUSBRE88N00L 20120924.

105 **"Life is hard for us":** Kane, *Haunted Empire,* 64.

105 **Prior to his death:** David Barboza, "iPhone Maker in China Is Under Fire After a Suicide," *New York Times,* July 27, 2009, accessed September 14, 2018, www.nytimes .com/2009/07/27/technology/companies/27apple.html?_r=2&scp=1&sq=foxconn &st=cse.

105 **But when the fourteenth:** Joel Johnson, "1 Million Workers. 90 Million iPhones. 17 Suicides. Who's to Blame?," *Wired,* February 28, 2011, accessed September 14, 2018, www.wired.com/2011/02/ff-joelinchina/.

106 **As a first step:** Frederik Balfour and Tim Culpan, "The Man Who Makes Your iPhone," Bloomberg, September 9, 2010, accessed September 14, 2018, www.bloom berg.com/news/articles/2010-09-09/the-man-who-makes-your-iphone.

106 **Finally, it set up:** John Vause, "Inside China Factory Hit by Suicides," CNN, June 2, 2010, accessed September 13, 2018, http://edition.cnn.com/2010/WORLD/asiapcf /06/01/china.foxconn.inside.factory/index.html.

106 **"How could a company":** Brent Schlender and Rick Tetzeli, *Becoming Steve Jobs: The Evolution of a Reckless Upstart into a Visionary Leader* (Toronto: Signal, 2016), 368.

106 **"We're all over this":** "Steve Jobs—Foxconn," YouTube, posted by pudox34, September 29, 2011, accessed September 14, 2018, www.youtube.com/watch?v=2gOu50HaEvs.

106 **The Fair Labor Association released:** Fair Labor Association, "Foxconn Verification Status Report," August 2012, www.fairlabor.org/sites/default/files/documents/reports/foxconn_verification_report_final.pdf.

107 **But reducing hours:** Fair Labor Association, "Foxconn Verification Status Report."

108 **"If they really wanted to change":** Author interview with Li Qiang, April 2018.

108 **"The scale of what":** Author interview with Ted Smith, April 2018.

109 **"I would like to see vertical":** Author interview with Jeff Ballinger, April 2018.

109 **"Workers should participate":** Author interview with Jenny Chan, April 2018.

109 **"If any of these corporations":** Author interview with Heather White, April 2018.

110 **"I actually feel like":** Author interview with Deirdre O'Brien, March 2018.

110 **"I hope that the Board will":** "Locke to Chair Apple's Academic Advisory Board," Watson Institute, Brown University, June 26, 2013, https://watson.brown.edu/news/2013/locke-chair-apples-academic-advisory-board.

111 **Each year, Apple has also:** An archive of reports is at the bottom of this page: www.apple.com/supplier-responsibility/.

111 **Jeff Williams said:** "Apple Releases 12th Annual Supplier Responsibility Report," Apple, March 7, 2018, www.apple.com/newsroom/2018/03/apple-releases-12th-annual-supplier-responsibility-progress-report/

111 **In 2018, Apple forced:** Stephen Nellis, "Apple Finds More Serious Supplier Problems as Its Audits Expand," Reuters, March 7, 2018, www.reuters.com/article/us-apple-suppliers/apple-finds-supplier-problems-as-its-audits-expand-idUSKCN1GK04G.

112 **"Pushing [this edict]":** Michael Rose, "Made for iPhone Manufacturers May Have to Comply with Apple's Supplier Responsibility Code," Engadget, November 8, 2012, www.engadget.com/2012/11/08/made-for-iphone-manufacturers-may-have-to-comply-with-apples-su/.

112 **"Apple's inclusion in the top tier":** *Electronic Industry Trends Report,* Baptist World Aid Australia, https://baptistworldaid.org.au/resources/ethical-electronics-guide/.

113 **Apple claimed that the number:** "iPhone 5 Pre-orders Top Two Million in First 24 Hours," Apple, September 17, 2012, accessed September 14, 2018, www.apple.com/newsroom/2012/09/17iPhone-5-Pre-Orders-Top-Two-Million-in-First-24-Hours/.

113 **"I'm incredibly proud":** Darrell Etherington, "Apple Posts a Video Remembering Steve Jobs and Highlighting His Greatest Achievements," TechCrunch, October 5, 2012, accessed September 14, 2018, https://techcrunch.com/2012/10/05/apple-posts-a-video-remember-steve-jobs-and-highlighting-his-greatest-achievements/.

113 **In December 2012:** "The World's 100 Most Influential People: 2012," *Time,* accessed September 14, 2018, http://content.time.com/time/specials/packages/completelist/0,29569,2111975,00.html.

113 **"It is difficult to imagine":** "The World's 100 Most Influential People: 2012."

114 **Analysts pointed out:** Jay Yarow, "Chart of the Day: Steve Jobs Leaves, Apple's Stock

Soars," Business Insider, September 20, 2011, accessed September 14, 2018, www
.businessinsider.com/chart-of-the-day-apple-stock-after-steve-jobs-2011-9.

114 **By the end of January 2012:** Leander Kahney, *Jony Ive: The Genius Behind Apple's Greatest Products* (New York: Portfolio, 2014).

Chapter 7: Finding His Feet with Hot New Products

115 **Despite its reporting:** "Apple Reports First Quarter Results," Apple, January 24, 2012, accessed September 14, 2018, www.apple.com/newsroom/2012/01/24Apple -Reports-First-Quarter-Results/.

115 **"I don't like it either":** Jill Treanor and Heidi Moore, "Apple's Share Price: Tim Cook Tells Angry Investors: 'I Don't Like It Either,'" *Guardian*, February 28, 2013, accessed September 14, 2018, www.theguardian.com/technology/2013/feb/27/apple-tim-cook -angry-investors-dont-like-it-either.

116 **"very important country":** Charles Riley, "Tim Cook: China Will Be Apple's Top Market," CNNMoney, January 11, 2013, accessed September 14, 2018, http://money .cnn.com/2013/01/11/technology/china-tim-cook-apple/index.html.

116 **Between 2010 and 2012:** Jared Newman, "Apple in China: By the Numbers," *Macworld*, November 5, 2013, accessed September 14, 2018, www.macworld.com/article /2056896/apple-in-china-by-the-numbers.html.

117 **"empty and self-praising":** Paul Mozur, "Beijing Takes Another Bite at Apple," *Wall Street Journal*, March 26, 2013, accessed September 14, 2018, www.wsj.com/articles /SB10001424127887323466204578382101284619638.

117 **"lack of communications":** Dominic Rushe, "Tim Cook Apologises After Chinese Media Rounds on Apple," *Guardian*, April 2, 2013, accessed September 14, 2018, www .theguardian.com/technology/2013/apr/01/apple-tim-cook-china-apology.

118 **"Apple wants to focus":** Connie Guglielmo, "Apple, Called a U.S. Tax Dodger, Says It's Paid 'Every Single Dollar' of Taxes Owed," *Forbes*, May 21, 2013, accessed September 14, 2018, www.forbes.com/sites/connieguglielmo/2013/05/21/apple-called-a-tax -dodger-by-senate-committee-apple-says-system-needs-to-be-dramatically-simplified /#240e47e53384.

118 **"all the taxes we owe":** Guglielmo, "Apple, Called a U.S. Tax Dodger."

118 **It was "very expensive":** Guglielmo, "Apple, Called a U.S. Tax Dodger."

119 **In December 2017:** Daisuke Wakabayashi and Brian X. Chen, "Apple, Capitalizing on New Tax Law, Plans to Bring Billions in Cash Back to the U.S.," *New York Times*, January 17, 2018, www.nytimes.com/2018/01/17/technology/apple-tax-bill-repatriate -cash.html.

119 **Apple was doing important:** Guglielmo, "Apple, Called a U.S. Tax Dodger."

119 **One, Senator Rand Paul:** Guglielmo, "Apple, Called a U.S. Tax Dodger."

120 **"thought it was very important":** Philip Elmer-Dewitt, "Apple's Tim Cook on Watches, Taxes, and How He's Like Steve Jobs," *Fortune*, May 29, 2013, accessed September 14,

2018, http://fortune.com/2013/05/29/apples-tim-cook-on-watches-taxes-and-how-hes-like-steve-jobs/.

120 **"frustrating for investors"**: Elmer-Dewitt, "Apple's Tim Cook on Watches, Taxes and How He's Like Steve Jobs."

120 **a "strong desire to set"**: "Preliminary Proxy Statement," SEC, accessed September 14, 2018, www.sec.gov/Archives/edgar/data/320193/000119312513486406/d648739dpre14a.htm.

120 **Even with the pay cut:** Ben Fox Rubin, "Cook Paid $4.2 Million," *Wall Street Journal*, December 28, 2012, accessed September 14, 2018, www.wsj.com/articles/SB10001424127887323530404578205430471522020.

120 **"not unprecedented"**: "Tim Cook Live at D11," The Verge, accessed September 14, 2018, https://live.theverge.com/tim-cook-d11-liveblog/.

120 **He reassured fans:** "Tim Cook Live at D11."

121 **"preternatural calm"**: "Apple's Tim Cook Leaves the D11 Audience Begging for . . . Anything," *Guardian*, www.theguardian.com/technology/blog/2013/jun/03/wall-street-journal-tech-conference.

121 **a "nice conversation"**: Carl Icahn, Twitter post, August 13, 2013, 11:25 a.m., https://twitter.com/Carl_C_Icahn/status/367351130776285184.

121 **"We currently have"**: Carl Icahn, Twitter post, August 13, 2013, 11:21 a.m., https://twitter.com/Carl_C_Icahn/status/367350206993399808.

122 **"We understood people"**: Marco Della Cava, "Jony Ive: The Man Behind Apple's Magic Curtain," *USA Today*, September 19, 2013, accessed September 14, 2018, www.usatoday.com/story/tech/2013/09/19/apple-jony-ive-craig-federighi/2834575/.

122 **Some labeled iOS 7:** Joshua Topolsky, "The Design of iOS 7: Simply Confusing," The Verge, June 10, 2013, accessed September 14, 2018, www.theverge.com/apple/2013/6/10/4416726/the-design-of-ios-7-simply-confusing.

122 **The A7's redesigned architecture:** Mario Aguilar, "iPhone A7 Chip Benchmarks: Forget the Specs, It Blows Everything Away," Gizmodo, October 4, 2013, accessed September 14, 2018, https://gizmodo.com/iphone-a7-chip-benchmarks-forget-the-specs-it-blows-e-1350717023.

123 **With a little help:** "First Weekend iPhone Sales Top Nine Million, Sets New Record," Apple, September 23, 2013, accessed September 14, 2018, www.apple.com/newsroom/2013/09/23First-Weekend-iPhone-Sales-Top-Nine-Million-Sets-New-Record/.

123 **The device even outsold:** "Apple Inc. (AAPL) iPhone 5S Outsold Galaxy S5: Best Selling Phone 2014," Dazeinfo, July 18, 2014, accessed September 14, 2018, https://dazeinfo.com/2014/07/18/apple-inc-aapl-iphone-5s-samsung-galaxy-s5-top-selling-smartphones-2014/.

123 **Though the beginning of 2013:** "Apple Reports First Quarter Results," Apple, January 27, 2014, accessed September 14, 2018, www.apple.com/newsroom/2014/01/27Apple-Reports-First-Quarter-Results/.

123 **"We are really happy":** Dominic Rushe, "Apple Shares Fall Despite Announcement of Record iPhone and iPad Sales," *Guardian*, January 27, 2014, accessed September 14, 2018, www.theguardian.com/technology/2014/jan/27/apple-shares-fall-record-iphone-ipad-sales.

123 **"As many of us prepare":** Rex Crum, "Tim Cook's '. . . One More Thing' Is All About Luck," *The Tell* (blog), Marketwatch, December 23, 2013, accessed September 14, 2018, http://blogs.marketwatch.com/thetell/2013/12/23/tim-cooks-one-more-thing-is-all-about-luck/.

124 **"a lot to look forward to":** Crum, "Tim Cook's '. . . One More Thing' Is All About Luck."

124 **"With health care, there is":** Serenity Caldwell, "This Is Tim: Cook Talks to Charlie Rose About Apple Watch, Samsung, and the Future," *Macworld*, September 16, 2014, accessed September 14, 2018, www.macworld.com/article/2684302/this-is-tim-cook-talks-to-charlie-rose-about-apple-watch-samsung-and-the-future.html.

125 **"I think if you had":** Samuel Gibbs, "Apple's Tim Cook: 'I Don't Want My Nephew on a Social Network,'" *Guardian*, January 19, 2018, accessed September 14, 2018, www.theguardian.com/technology/2018/jan/19/tim-cook-i-dont-want-my-nephew-on-a-social-network.

125 **"Coding gives people":** Jasper Hamill, "Apple CEO Tim Cook Reveals How YOU Can Follow in His Footsteps," *The Sun*, October 13, 2017, accessed September 13, 2018, www.thesun.co.uk/tech/4663185/apple-ceo-tim-cook-reveals-a-big-career-secret-and-tells-how-you-can-follow-in-his-footsteps/.

125 **"We try to continually":** David Phelan, "Apple CEO Tim Cook Speaks Out on Brexit," *Independent*, February 10, 2017, accessed September 14, 2018, www.independent.co.uk/life-style/gadgets-and-tech/features/apple-tim-cook-boss-brexit-uk-theresa-may-number-10-interview-ustwo-a7574086.html.

126 **In October 2013:** "Angela Ahrendts to Join Apple as Senior Vice President of Retail and Online Stores," Apple, October 14, 2013, accessed September 14, 2018, www.apple.com/newsroom/2013/10/15Angela-Ahrendts-to-Join-Apple-as-Senior-Vice-President-of-Retail-and-Online-Stores/.

126 **"not a techie":** Anieze Osakwe, "Apple Exec Angela Ahrendts Recalls Telling Tim Cook, 'I'm Not a Techie,' in First Meeting," ABC News, January 9, 2018, accessed September 14, 2018, https://abcnews.go.com/Business/apple-exec-angela-ahrendts-recalls-telling-tim-cook/story?id=52222468.

126 **"values and our focus":** Mark Gurman, "Tim Cook Talks Hiring of Angela Ahrendts as Retail Chief, Says She Is 'Best Person in the World for This Role,'" 9to5Mac, October 15, 2013, accessed September 14, 2018, https://9to5mac.com/2013/10/15/tim-cook-talks-hiring-of-angela-ahrendts-as-retail-chief-says-she-is-best-person-in-the-world-for-this-role/.

126 **Ron Johnson, former:** Adam Satariano and Adam Ewing, "Apple Hires Burberry Chief to End Search for Retail Head," Bloomberg, October 15, 2013, accessed Sep-

tember 14, 2018, www.bloomberg.com/news/articles/2013-10-15/apple-names-burberry
-chief-ahrendts-head-of-retail-operations.

127 **"We think of Apple retail":** Abha Bhattarai, "Apple Wants Its Stores to Become 'Town Squares.' But Skeptics Call It a 'Branding Fantasy,'" *Washington Post,* September 13, 2017, accessed September 14, 2018, www.washingtonpost.com/news/business/wp /2017/09/13/apple-wants-its-stores-to-become-town-squares-but-skeptics-call-it-a -branding-fantasy/?utm_term=.6bf6b8aba249.

127 **Williams has been called:** JP Mangalindan, "Jeff Williams: Apple's Other Operations Whiz," *Fortune,* June 20, 2014, accessed September 14, 2018, http://fortune .com/2011/09/13/jeff-williams-apples-other-operations-whiz/.

127 **"Tall, lean, and gray-haired":** Adam Lashinsky, *Inside Apple* (Wiley, 2012), 107.

128 **"With Jeff, what you see":** Mangalindan, "Jeff Williams: Apple's Other Operations Whiz."

128 **He wasn't mentioned once:** Walter Isaacson, *Steve Jobs* (New York: Simon & Schuster, 2011).

128 **He also helped:** Mangalindan, "Jeff Williams: Apple's Other Operations Whiz."

128 **"Jeff Williams is doing":** "Jeff Williams: Apple CEO Material," Above Avalon, accessed September 14, 2018, www.aboveavalon.com/notes/2015/1/29/jeff-williams -apple-ceo-material.

129 **In May 2014, Apple announced:** "Apple to Acquire Beats Music & Beats Electronics," Apple, May 28, 2014, accessed September 14, 2018, www.apple.com/newsroom /2014/05/28Apple-to-Acquire-Beats-Music-Beats-Electronics/.

129 **During an interview:** "Tim Cook," *Charlie Rose,* September 12, 2014, accessed September 14, 2018, https://charlierose.com/videos/18663.

130 **"What Beats brings":** Peter Kafka, "Tim Cook Explains Why Apple Is Buying Beats (Q&A)," Recode, May 28, 2014, accessed September 14, 2018, www.recode.net/2014 /5/28/11627398/tim-cook-explains-why-apple-is-buying-beats-qa.

130 **Dre was largely absent:** "Apple—WWDC 2015," YouTube, posted by Apple, June 15, 2015, accessed September 14, 2018, www.youtube.com/watch?v=_p8AsQhaVKI.

130 **A year later:** Kara Swisher, "Bozoma Saint John Is Leaving Uber for Endeavor," Recode, June 11, 2018, accessed September 14, 2018, www.recode.net/2018/6/11/17449978 /bozoma-saint-john-depart-uber-endeavor.

131 **Apple Music has become:** Stephen Nellis, "Apple Music Hits 38 Million Paid Subscribers," Reuters, March 12, 2018, accessed September 14, 2018, www.reuters.com /article/us-apple-music/apple-music-hits-38-million-paid-subscribers -idUSKCN1GO2G2.

131 **"market-leading strengths":** "Apple and IBM Forge Global Partnership to Transform Enterprise Mobility," Apple, July 15, 2014, accessed September 14, 2018, www .apple.com/newsroom/2014/07/15Apple-and-IBM-Forge-Global-Partnership-to -Transform-Enterprise-Mobility/.

131 **"This is an area":** "Tim Cook on Apple TV (Sept. 12, 2014) | Charlie Rose Show,"

YouTube, posted by Charlie Rose, September 12, 2014, accessed September 14, 2018, www.youtube.com/watch?v=oBMo8Oz9jsQ.

132 **Apple shares climbed 2.59 percent:** "Apple, IBM Shares Up After Deal Announcement; BBRY Down," CNBC, July 15, 2014, www.cnbc.com/2014/07/15/apple-ibm -shares-up-after-deal-annoucement-bbry-down.html.

132 **Shares in BlackBerry:** Euan Rocha, "Apple-IBM Deal Dents BlackBerry's Prospects, Slams Stock," Reuters, July 16, 2014, www.reuters.com/article/us-blackberry-stocks /apple-ibm-deal-dents-blackberrys-prospects-slams-stock-idUSKBN0FL1MZ20140716.

132 **More than thirty-eight hundred:** Bill Siwicki, "3 Years In, Here's What the Apple and IBM Partnership Has Achieved," *Healthcare IT News*, July 10, 2017, www .healthcareitnews.com/news/3-years-heres-what-apple-and-ibm-mobilefirst-ios -partnership-has-achieved.

132 **"the biggest advancement":** "Tim Cook on Apple TV (Sept. 12, 2014) | Charlie Rose Show."

133 **The radically redesigned:** "Apple Announces Record Pre-orders for iPhone 6 & iPhone 6 Plus Top Four Million in First 24 Hours," Apple, September 15, 2014, accessed September 14, 2018, www.apple.com/newsroom/2014/09/15Apple-Announces-Record -Pre-orders-for-iPhone-6-iPhone-6-Plus-Top-Four-Million-in-First-24-Hours/.

133 **more than ten million sold:** "First Weekend iPhone Sales Top 10 Million, Set New Record," Apple, September 22, 2014, accessed September 14, 2018, www.apple.com /newsroom/2014/09/22First-Weekend-iPhone-Sales-Top-10-Million-Set-New -Record/.

133 **A YouTube video:** "iPhone 6 Plus Bend Test," YouTube, posted by Unbox Therapy, September 23, 2014, accessed September 14, 2018, www.youtube.com/watch?v =znK652H6yQM.

133 **"high quality standard":** Jeremy Horwitz, "Apple Knew iPhone Bent Easily, but Released It and Downplayed Issues," VentureBeat, May 24, 2018, accessed September 14, 2018, https://venturebeat.com/2018/05/24/apple-knew-iphone-6-bent-easily -but-released-it-and-downplayed-issues/.

134 **Assuming no abuse:** Kif Leswing, "Leaked Document Shows How Apple Decides to Replace or Repair Your iPhone," Business Insider, September 1, 2017, accessed September 14, 2018, www.businessinsider.com/leaked-apple-document-how-geniuses-decide -replace-repair-iphones-warranty-2017-9.

134 **Before Bendgate had a chance:** Mike Beasley, "Apple Releases iOS 8.0.2 to Address Cellular and Touch ID Issues in Previous Update," 9to5Mac, September 26, 2014, accessed September 14, 2018, https://9to5mac.com/2014/09/25/apple-releases-ios -8-0-2-to-address-cellular-and-touch-id-issues-in-previous-update/.

134 **"We are actively investigating":** Chris Welch, "Apple Pulls iOS 8.0.1 After Users Report Major Problems with Update," The Verge, September 24, 2014, accessed September 14, 2018, www.theverge.com/2014/9/24/6839235/apple-ios-8-0-1-released.

134 **Data from Kantar:** Charles Arthur, "Impatient iPhone Users Switching to Larger-

Screened Samsung Galaxy S5," *Guardian*, June 30, 2014, accessed September 14, 2018, www.theguardian.com/technology/2014/jun/30/impatient-iphone-users-switching -to-larger-screened-samsung-galaxy-s5.

135 **Combined, the new handsets:** Neil Hughes, "iPhone 5s Represents 3.8% of All iPhones in Use, US Has Highest 5c Adoption Rate," AppleInsider, October 28, 2013, accessed September 14, 2018, https://appleinsider.com/articles/13/10/28/iphone-5s -represents-38-of-all-iphones-in-use-us-has-highest-5c-adoption-rate.

135 **"a higher rate of switchers":** Jim Edwards, "Tim Cook Said This Word 5 Times on Apple's Earnings Call Last Night—Here's Why It's So Important," Business Insider, April 28, 2015, accessed September 14, 2018, www.businessinsider.com/apple-ceo -tim-cook-talks-about-android-switchers-2015-4.

135 **"Most people who have worked":** Jason Del Rey, "Apple Introduces Apple Pay to Try to Replace Your Wallet," Recode, September 9, 2014, accessed September 14, 2018, www.recode.net/2014/9/9/11630686/apple-introduces-apple-pay-to-try-to-replace -your-wallet.

135 **"You are not our product":** Serenity Caldwell, "This Is Tim: Cook at the Goldman Sachs Conference," iMore, February 10, 2015, accessed September 14, 2018, www .imore.com/tim-cook-goldman-sachs-conference.

136 **During an appearance:** Alex Wilhelm, "Apple CEO Tim Cook: Apple Pay Activated 1M Cards in 72 Hours," TechCrunch, October 27, 2014, accessed September 14, 2018, https://techcrunch.com/2014/10/27/apple-ceo-tim-cook-apple-pay-activated-1m -cards-in-72-hours/.

136 **"more than the total":** Nathan Ingraham, "Tim Cook Says That Apple Pay Is Already the Leader in Contactless Payments," The Verge, October 28, 2014, accessed September 14, 2018, www.theverge.com/2014/10/27/7082013/tim-cook-says-that-apple -pay-is-already-the-leader-in-contactless-payments.

136 **"I'm unbelievably shocked":** Mark Sullivan, "Tim Cook: '2015 Will Be the Year of Apple Pay,'" VentureBeat, January 27, 2015, accessed September 14, 2018, https:// venturebeat.com/2015/01/27/tim-cook-2015-will-be-the-year-of-apple-pay/.

136 **In June 2016:** Lena Rao, "Grubhub Adds Apple Pay to Food Delivery Apps," *Fortune*, June 2, 2016, accessed September 14, 2018, http://fortune.com/2016/06/02/grubhub -apple-pay/.

137 **"taken off slower":** Steven Anderson, "Apple's Tim Cook Surprised at Pace of Mobile Payments Adoption," *Payment Week*, February 16, 2018, accessed September 14, 2018, https://paymentweek.com/2018-2-16-apples-tim-cook-surprised-pace-mobile -payments-adoption/.

137 **"I'm hoping to be alive":** Kif, Twitter post, February 13, 2018, 9:52 a.m., https:// twitter.com/kifleswing/status/963471023843651584.

137 **"well over a billion":** Jeff Gamet, "Apple Pay Transactions Top 1 Billion in Q3 2018," Mac Observer, July 31, 2018, accessed September 14, 2018, www.macobserver.com /news/apple-pay-transactions-top-1-billion-q3-2018/.

137 **"the next chapter"**: Andrea Chang, "Apple Watch Unveiling Shows CEO Tim Cook's Time Has Come," *Los Angeles Times*, September 9, 2014, accessed September 14, 2018, www.latimes.com/business/la-fi-apple-cook-20140910-story.html.

138 **"a precise timepiece"**: "Apple Special Event 2014—Apple Watch," YouTube, posted by the unofficial AppleKeynotes channel, September 10, 2014, accessed September 14, 2018, www.youtube.com/watch?v=bdyVH5LqneU.

138 **Apple received "overwhelmingly positive"**: Juli Clover, "Apple CEO Tim Cook: We're 'Working Hard' to Remedy Apple Watch Supply/Demand Imbalance," MacRumors, April 27, 2015, accessed September 14, 2018, www.macrumors.com/2015/04/27/apple-watch-supply-demand-balance/.

138 **It is believed that it took:** Arjun Kharpal, "Tim Cook: Health Care Opportunity 'Enormous,'" CNBC, May 24, 2016, accessed September 14, 2018, www.cnbc.com/2016/05/24/tim-cook-why-the-apple-watch-is-key-in-the-enormous-health-care-market.html.

138 **"We didn't talk about watches"**: Benjamin Clymer, "Apple, Influence, and Ive," *Hodinkee Magazine*, accessed September 14, 2018, www.hodinkee.com/magazine/jony-ive-apple.

138 **"The first discussion"**: Clymer, "Apple, Influence, and Ive."

139 **By the end of November 2014:** "Apple Now Worth a Whopping $700 Billion," CNN, https://money.cnn.com/2014/11/25/investing/apple-700-billion/index.html.

Chapter 8: A Greener Apple

141 **"Sustainability was pretty much"**: Luke Dormehl, "Why Tim Cook's Green Push Goes Back to Apple's Roots," Cult of Mac, July 26, 2015, accessed September 14, 2018, www.cultofmac.com/275699/apple-green-campaign.

142 **"Pure looks trumped"**: Author interview with Abraham Farag, August 2014.

142 **The analysis found**: David Santillo et al., *Missed Call: iPhone's Hazardous Chemicals* (Amsterdam: Greenpeace International, October 2007), www.greenpeace.org/archive-international/PageFiles/25275/iPhones-hazardous-chemicals.pdf.

142 **"If Apple really wants"**: Santillo et al., *Missed Call*.

143 **"Apple is ahead of"**: Gregg Keizer, "Steve Jobs Promises 'Greener Apple,'" *Computerworld*, May 3, 2007, www.computerworld.com/article/2544865/computer-hardware/steve-jobs-promises--greener-apple-.html.

143 **"There's no excuse"**: "Steve Jobs Greener Apple Update," Greenpeace International, July 5, 2015, accessed September 14, 2018, www.greenpeace.org/usa/steve-jobs-greener-apple-update/.

143 **"unreasonably high threshold"**: Greenpeace International, *Guide to Greener Electronics*, June 2009, www.greenpeace.org/usa/wp-content/uploads/legacy/Global/usa/report/2009/7/guide-to-greener-electronics-12.pdf.

143 **It was also awarded:** Greenpeace International, *Guide to Greener Electronics*.

144 **A January 2011 study:** Jonathan Watts, "Apple Secretive About 'Polluting and Poi-

soning' Supply Chain, Says Report," *Guardian*, January 20, 2011, accessed September 14, 2018, www.theguardian.com/environment/2011/jan/20/apple-pollution-supply-chain.

144 **An accompanying video:** Watts, "Apple Secretive About 'Polluting and Poisoning' Supply Chain."

144 **The poisonings happened:** Watts, "Apple Secretive About 'Polluting and Poisoning' Supply Chain."

145 **"This attitude means":** Watts, "Apple Secretive About 'Polluting and Poisoning' Supply Chain."

145 **Ma, a former investigative journalist:** Friends of Nature, IPE, Green Beagle, *The Other Side of Apple*, February 20, 2011, www.business-humanrights.org/sites/default/files/media/documents/it_report_phase_iv-the_other_side_of_apple-final.pdf.

145 **Despite making a significant:** "The Other Side of Apple II. Pollution Spreads Through Apple's Supply Chain," GoodElectronics, September 1, 2011, accessed September 14, 2018, https://goodelectronics.org/the-other-side-of-apple-ii-pollution-spreads-through-apples-supply-chain/.

145 **In a forty-six-page:** "The Other Side of Apple II." Also see www.ipe.org.cn/Upload/Report-IT-V-Apple-II.pdf.

145 **"Apple is committed":** David Barboza, "Apple Cited as Adding to Pollution in China," *New York Times*, September 1, 2011, accessed September 14, 2018, www.nytimes.com/2011/09/02/technology/apple-suppliers-causing-environmental-problems-chinese-group-says.html.

146 **"build a relationship":** Yukari Iwatani Kane, *Haunted Empire: Apple After Steve Jobs* (London: William Collins, 2015), 123.

147 **"The meeting lasted":** Kane, *Haunted Empire*, 123.

147 **Apple was labeled:** Greenpeace, *How Dirty Is Your Data?*, www.greenpeace.org/archive-international/Global/international/publications/climate/2011/Cool%20IT/dirty-data-report-greenpeace.pdf.

148 **With no response:** C. J. Hughes, "Greenpeace Protests Apple Energy Practices by Releasing Balloons," April 24, 2012, https://cityroom.blogs.nytimes.com/2012/04/24/greenpeace-protests-apples-energy-practices-by-releasing-balloons/.

149 **"Apple's announcement today":** Robert McMillan, "After Greenpeace Protests, Apple Promises to Dump Coal Power," *Wired*, June 3, 2017, accessed September 14, 2018, www.wired.com/2012/05/apple-coal/.

149 **"Only then will customers":** McMillan, "After Greenpeace Protests, Apple Promises to Dump Coal Power."

149 **Only eighteen months before:** "Apple Supplier Facing 'Harshest' Pollution Penalty," Shanghai.gov, February 22, 2013, accessed September 14, 2018, www.shanghai.gov.cn/shanghai/node27118/node27818/u22ai70987.html.

150 **"Jackson can make Apple":** David Price, "Why Apple Was Bad for the Environment (and Why That's Changing)," *Macworld UK*, January 3, 2017, accessed September

14, 2018, www.macworld.co.uk/feature/apple/complete-guide-apples-environmental-impact-green-policies-3450263/.

150 **"Tim obviously is very interested":** Author interview with Lisa Jackson, March 2018.

151 **Greenpeace's Gary Cook:** Author interview with Gary Cook, March 2018.

152 **"Whether it's improving":** "Tim Cook Wants Apple to Be a 'Force for Good,'" Cult of Mac, July 26, 2015, accessed September 14, 2018, www.cultofmac.com/251795/tim-cook-wants-apple-to-be-a-force-for-good/.

152 **In 2014, Apple was:** Elsa Wenzel, "Apple, Facebook, Google Score in Greenpeace Data Center Ratings," GreenBiz, April 2, 2014, accessed September 14, 2018, www.greenbiz.com/blog/2014/04/02/google-apple-facebook-good-greenpeace-energy-ratings.

152 **"a trade-off between":** Darrell Etherington, "Apple CEO Tim Cook Says Tech Companies Should Accept No Compromises on Climate Change Issues," TechCrunch, September 22, 2014, accessed September 14, 2018, https://techcrunch.com/2014/09/22/apple-ceo-tim-cook-on-climate-change/.

152 **By 2014, Apple managed:** Apple Inc., *Environmental Responsibility Report,* July 2014, www.apple.com/environment/reports/docs/apple_environmental_responsibility_report_0714.pdf.

153 **"were burning cords":** Author interview with Lisa Jackson, March 2018.

153 **In February 2015:** Jacob Pramuk, "Apple to Build $850M Solar Energy Farm in CA," CNBC, February 11, 2015, accessed September 14, 2018, www.cnbc.com/2015/02/10/apple-ceo-tim-cook-will-partner-with-first-solar-on-850m-ca-solar-farm.html.

153 **"Apple is leading the way":** "First Solar and Apple Strike Industry's Largest Commercial Power Deal," *Business Wire,* www.businesswire.com/news/home/20150210006559/en/Solar-Apple-Strike-Industry's-Largest-Commercial-Power.

153 **"Apple's commitment was instrumental":** Maggie McGrath, "First Solar Jumps on $850 Million Partnership with Apple," *Forbes,* February 10, 2015, accessed September 14, 2018, www.forbes.com/sites/maggiemcgrath/2015/02/10/first-solar-jumps-on-850-million-investment-from-apple/&refURL=&referrer=#41d0a09ead1e.

154 **"the time for talk":** Christina Farr, "Apple Investing $850 Million in California Solar Farm," Reuters, February 10, 2015, accessed September 14, 2018, www.reuters.com/article/us-apple-cook/apple-investing-850-million-in-california-solar-farm-idUSKBN0LE2RN20150210.

154 **In October 2015:** "Apple Launches New Clean Energy Programs in China to Promote Low-Carbon Manufacturing and Green Growth," Apple, October 22, 2015, accessed September 14, 2018, www.apple.com/newsroom/2015/10/22Apple-Launches-New-Clean-Energy-Programs-in-China-To-Promote-Low-Carbon-Manufacturing-and-Green-Growth/.

154 **Foxconn, one of Apple's:** Valerie Volcovici, "Apple to Build More Solar Projects in China, Green Its Suppliers," Reuters, October 22, 2015, accessed September 14, 2018,

www.reuters.com/article/us-apple-renewables-china/apple-to-build-more-solar-projects-in-china-green-its-suppliers-idUSKCN0SG02V20151022.

154 **In an interview:** Jena McGregor, "Tim Cook, the Interview: Running Apple 'Is Sort of a Lonely Job,'" *Washington Post,* August 13, 2016, accessed September 14, 2018, www.washingtonpost.com/sf/business/2016/08/13/tim-cook-the-interview-running-apple-is-sort-of-a-lonely-job/.

155 **Then, in December 2016:** He Wei and Liu Zheng, "Apple Reaches Clean Energy Deal with Goldwind," *China Daily,* accessed September 14, 2018, www.chinadaily.com.cn/business/tech/2016-12/09/content_27618014.htm.

155 **The deal appeared:** Ivan Shumkov, "Goldwind Revises Down Profit Guidance, Blames Deal with Apple," Renewables Now, accessed September 14, 2018, https://renewablesnow.com/news/goldwind-revises-down-profit-guidance-blames-deal-with-apple-576089/.

155 **But a year later:** "Apple Launches New Clean Energy Fund in China," Apple, July 12, 2018, accessed September 14, 2018, www.apple.com/newsroom/2018/07/apple-launches-new-clean-energy-fund-in-china/.

155 **The China Clean Energy Fund:** "Apple Launches $300 Mn 'Green' Fund for China Suppliers," Phys.org, July 2018, accessed September 14, 2018, https://phys.org/news/2018-07-apple-mn-green-fund-china.html.

155 **On Earth Day:** "Apple Now Globally Powered by 100 Percent Renewable Energy," Apple, April 9, 2018, accessed September 14, 2018, www.apple.com/newsroom/2018/04/apple-now-globally-powered-by-100-percent-renewable-energy/.

156 **"We're really adamant":** Author interview with Lisa Jackson, March 2018.

156 **Up to 77 percent:** Apple Inc., *Environmental Responsibility Report,* July 2018, www.apple.com/environment/pdf/Apple_Environmental_Responsibility_Report_2018.pdf.

157 **"It may take a few years":** Author interview with Gary Cook, March 2018.

157 **In July 2018:** "Samsung Electronics to Expand Use of Renewable Energy," Samsung Global Newsroom, June 14, 2018, accessed September 14, 2018, https://news.samsung.com/global/samsung-electronics-to-expand-use-of-renewable-energy.

157 **"Amazon is just going":** Author interview with Gary Cook, March 2018.

158 **"It sounds crazy":** Apple Inc., *Environmental Responsibility Report,* July 2017, https://images.apple.com/environment/pdf/Apple_Environmental_Responsibility_Report_2017.pdf.

158 **There are currently:** Kif Leswing, "Apple's iPhone-Destroying Robots Are 'Operating' in California and Europe," Business Insider, March 11, 2017, accessed September 14, 2018, www.businessinsider.com/apples-iphone-robot-liam-update-2017-3.

158 **"Traditional e-waste recycling":** Charissa Rujanavech et al., "Liam—An Innovation Story," September 2016, accessed September 13, 2018, www.apple.com/environment/pdf/Liam_white_paper_Sept2016.pdf.

158 **Called Daisy, the robot:** "Apple Adds Earth Day Donations to Trade-in and Recycling

Program," Apple, April 19, 2018, accessed September 14, 2018, www.apple.com
/newsroom/2018/04/apple-adds-earth-day-donations-to-trade-in-and-recycling
-program/.

159 **Currently, the percentage:** Apple Inc., *Environmental Responsibility Report,* July
2018.

159 **"devoted entirely to getting":** Author interview with Lisa Jackson, March 2018.

159 **Apple also announced:** "Apple Paves the Way for Breakthrough Carbon-Free Aluminum Smelting Method," Apple, May 10, 2018, accessed September 14, 2018, www
.apple.com/newsroom/2018/05/apple-paves-the-way-for-breakthrough-carbon-free
-aluminum-smelting-method/.

160 **"We are proud to be part":** "Apple Paves the Way for Breakthrough Carbon-Free
Aluminum Smelting Method."

160 **As of 2017:** "Apple Releases 12th Annual Supplier Responsibility Progress Report,"
Apple, March 8, 2018, accessed September 14, 2018, www.apple.com/sg/newsroom
/2018/03/apple-releases-12th-annual-supplier-responsibility-progress-report.

160 **In an April 2016 Greenpeace report:** Gary Cook, "Greenpeace Responds to Apple
Environmental Progress Report," Greenpeace International, April 14, 2016, accessed
September 14, 2018, www.greenpeace.org/usa/news/greenpeace-responds-to-apple
-environmental-progress-report/.

161 **In the United States:** Apple Inc., *Environmental Responsibility Report,* July 2018.

161 **At "other companies":** Author interview with Gary Cook, March 2018.

162 **"We're very protective":** Author interview with Lisa Jackson, March 2018.

162 **"He doesn't let us just":** Author interview with Lisa Jackson, March 2018.

Chapter 9: Cook Fights the Law, and Wins

163 **Protecting the privacy:** "Exclusive: Brian Williams Interviews Apple CEO," NBC
News, www.nbcnews.com/video/exclusive-brian-williams-interviews-apple-ceo
-11421251878?v=railb&.

164 **Activation Lock makes:** Brian X. Chen, "Smartphones Embracing 'Kill Switches'
as Theft Defense," *New York Times,* June 19, 2014, accessed September 14, 2018, https://
bits.blogs.nytimes.com/2014/06/19/antitheft-technology-led-to-a-dip-in-iphone
-thefts-in-some-cities-police-say/.

165 **In November 2013:** Apple Inc., *Report on Government Information Requests,* November
5, 2013, www.apple.com/legal/privacy/transparency/requests-20131105-en.pdf.

165 **iAds "sticks to the same":** Chris Smith, "This Might Be Apple CEO Tim Cook's
Most Important Message Yet," BGR, September 18, 2014, accessed September 14,
2018, https://bgr.com/2014/09/18/tim-cook-on-apple-privacy-2/.

166 **During his speech:** Farhad Manjoo, "What Apple's Tim Cook Overlooked in His
Defense of Privacy," *New York Times,* December 21, 2017, accessed September 14,
2018, www.nytimes.com/2015/06/11/technology/what-apples-tim-cook-overlooked
-in-his-defense-of-privacy.html.

166 **"Like many of you"**: Matthew Panzarino, "Apple's Tim Cook Delivers Blistering Speech on Encryption, Privacy," TechCrunch, June 2, 2015, accessed September 14, 2018, https://techcrunch.com/2015/06/02/apples-tim-cook-delivers-blistering-speech -on-encryption-privacy/.

167 **"I'm speaking to you"**: Panzarino, "Apple's Tim Cook Delivers Blistering Speech on Encryption, Privacy."

167 **"We don't think you"**: Panzarino, "Apple's Tim Cook Delivers Blistering Speech on Encryption, Privacy."

168 **Cook admitted during**: "What's Next for Apple?," CBS News, December 20, 2015, accessed September 14, 2018, www.cbsnews.com/news/60-minutes-apple-tim-cook -charlie-rose/.

168 **One of the first public discussions**: Jay Hathaway, "The NSA Has Nearly Complete Access to Apple's iPhone," Daily Dot, March 8, 2017, accessed September 14, 2018, www .dailydot.com/layer8/nsa-backdoor-iphone-access-camera-mic-appelbaum/.

168 **"Apple has never"**: Matthew Panzarino, "Apple Says It Has Never Worked with NSA to Create iPhone Backdoors, Is Unaware of Alleged DROPOUTJEEP Snooping Program," TechCrunch, December 31, 2013, accessed September 14, 2018, https:// techcrunch.com/2013/12/31/apple-says-it-has-never-worked-with-nsa-to-create -iphone-backdoors-is-unaware-of-alleged-dropoutjeep-snooping-program/.

169 **An earlier white paper**: Fred Raynal, "iMessage Privacy," *Quarkslab's Blog*, October 17, 2013, accessed September 14, 2018, https://blog.quarkslab.com/imessage-privacy .html.

169 **"discussed theoretical vulnerabilities"**: John Paczkowski, "Apple: No, We Can't Read Your iMessages (and We Don't Want To, Either)," AllThingsD, October 18, 2013, accessed September 14, 2018, http://allthingsd.com/20131018/apple-no-we-cant-read -your-imessages.

169 **"We believe that our customers"**: Alex Heath, "Apple Exposes Governments' Re- quests for Customer Data, Pushes for Greater Transparency in New Report," Cult of Mac, November 5, 2013, accessed September 14, 2018, www.cultofmac.com/253020 /apple-exposes-governments-requests-for-customer-data-in-new-report/.

169 **"no backdoor is a must"**: Kia Kokalitcheva, "Apple CEO Tim Cook Says No to NSA Accessing User Data," *Fortune*, October 20, 2015, accessed September 14, 2018, http:// fortune.com/2015/10/20/tim-cook-against-backdoor/.

170 **"None of the cases"**: Steve Kovach, "We Still Don't Have Assurance from Apple That iCloud Is Safe," Business Insider, September 2, 2014, accessed September 14, 2018, www.businessinsider.com/apple-statement-on-icloud-hack-2014-9.

170 **iCloud "wasn't hacked"**: Serenity Caldwell, "This Is Tim: Cook Talks to Charlie Rose About Apple Watch, Samsung, and the Future," *Macworld*, September 16, 2014, accessed September 14, 2018, www.macworld.com/article/2684302/this -is-tim-cook-talks-to-charlie-rose-about-apple-watch-samsung-and-the-future .html?page=2.

171 **"When I step back"**: Daisuke Wakabayashi, "Tim Cook Says Apple to Add Security

Alerts for iCloud Users," *Wall Street Journal*, September 5, 2014, accessed September 14, 2018, www.wsj.com/articles/tim-cook-says-apple-to-add-security-alerts-for-icloud -users-1409880977.

171 **Apple began implementing:** Eric Slivka, "Apple Now Sending Alert Emails When iCloud Accounts Accessed via Web," MacRumors, September 8, 2014, accessed September 14, 2018, www.macrumors.com/2014/09/08/icloud-alert-emails-web/.

172 **But Apple had refused:** Amanda Holpuch, "Tim Cook Says Apple's Refusal to Unlock iPhone for FBI Is a 'Civil Liberties' Issue," *Guardian*, February 22, 2016, accessed September 14, 2018, www.theguardian.com/technology/2016/feb/22/tim-cook-apple -refusal-unlock-iphone-fbi-civil-liberties.

172 **Apple was given:** Author interview with Bruce Sewell, March 2018.

172 **"was not a simple request":** Author interview with Bruce Sewell, March 2018.

173 **"is a famous":** Author interview with Bruce Sewell, March 2018.

174 **At 4:30 a.m.:** "A Message to Our Customers," Apple, February 16, 2016, accessed September 14, 2018, https://www.apple.com/customer-letter/.

175 **"Unlike our competitors":** "Privacy," Apple, accessed September 14, 2018, www .apple.com/privacy/government-information-requests/.

175 **"When the FBI filed":** Author interview with Bruce Sewell, March 2018.

176 **Cook, Sewell, and others:** Author interview with Bruce Sewell, March 2018.

176 **Once a backdoor had been created:** Author interview with Bruce Sewell, March 2018.

176 **"There was a sense":** Author interview with Bruce Sewell, March 2018.

177 **Apple's response drew:** Sam Thielman and Danny Yadron, "Crunch Time for Apple as It Prepares for Face-off with FBI," *Guardian*, February 27, 2016, accessed September 14, 2018, www.theguardian.com/technology/2016/feb/27/apple-fbi-congressional -hearing-iphone-encryption.

177 **"Despite having a court order":** "Apple's Line in the Sand Was Over a Year in the Making," *New York Times*, www.nytimes.com/2016/02/19/technology/a-yearlong -road-to-a-standoff-with-the-fbi.html.

177 **A few days later, Donald Trump:** "Donald Trump Boycott Apple If They Don't Help FBI," YouTube, February 20, 2016, accessed September 14, 2018, www.youtube.com /watch?v=xG7aus4ldxA.

177 **A Pew survey found:** Shiva Maniam, "More Support for Justice Department Than for Apple in Dispute over Unlocking iPhone," Pew Research Center for the People and the Press, February 22, 2016, accessed September 14, 2018, www.people-press .org/2016/02/22/more-support-for-justice-department-than-for-apple-in-dispute -over-unlocking-iphone/.

178 **According to that poll:** Jim Finkle, "Solid Support for Apple in iPhone Encryption Fight: Poll," Reuters, February 24, 2016, accessed September 14, 2018, www.reuters .com/article/us-apple-encryption-poll-idUSKCN0VX159.

178 **The difference was attributed:** Krishnadev Calamur, "Public Opinion Supports

Apple over the FBI—or Does It?," *Atlantic*, February 24, 2016, accessed September 14, 2018, www.theatlantic.com/national/archive/2016/02/apple-fbi-polls/470736/.

178 **By analyzing positive:** Gabrielle Hughes, "Social Media's Response to Apple vs. the FBI," Convince and Convert, accessed September 14, 2018, www.convinceandconvert .com/realtime-today/social-medias-response-to-apple-vs-the-fbi/.

178 **Several high-profile figures:** Stephen Foley and Tim Bradshaw, "Gates Breaks Ranks over FBI Apple Request," *Financial Times*, February 23, 2016, accessed September 14, 2018, www.ft.com/content/3559f46e-d9c5-11e5-98fd-06d75973fe09.

178 **In an editorial titled:** "Why Apple Is Right to Challenge an Order to Help the F.B.I.," *New York Times*, February 19, 2016, accessed September 14, 2018, www .nytimes.com/2016/02/19/opinion/why-apple-is-right-to-challenge-an-order-to -help-the-fbi.html.

179 **For the next two months:** Author interview with Bruce Sewell, March 2018.

179 **One PR rep said:** Author interview with Apple PR staffer, who asked to remain anonymous, March 2018.

179 **"This case is about":** "Tim Cook Says Apple's Refusal to Unlock iPhone for FBI Is a 'Civil Liberties' Issue," *Guardian*, www.theguardian.com/technology/2016/feb/22 /tim-cook-apple-refusal-unlock-iphone-fbi-civil-liberties.

179 **"I think a lot of reporters":** Author interview with Apple PR staffer, who asked to remain anonymous, March 2018.

180 **It was the "most important":** Jena McGregor, "Tim Cook's Interview About Apple's Fight with the FBI May Be the Most Important of His Career," *Washington Post*, February 26, 2016, accessed September 14, 2018, www.washingtonpost.com/news /on-leadership/wp/2016/02/26/tim-cooks-interview-about-apples-fight-with-the -fbi-may-be-the-most-important-of-his-career/.

180 **"This is not a rapacious":** Author interview with Bruce Sewell, March 2018.

180 **"The implications of the government's":** Spencer Ackerman, Sam Thielman, and Danny Yadron, "Apple Case: Judge Rejects FBI Request for Access to Drug Dealer's iPhone," *Guardian*, February 29, 2016, accessed September 14, 2018, www.theguard ian.com/technology/2016/feb/29/apple-fbi-case-drug-dealer-iphone-jun-feng-san -bernardino.

181 **"For us it was very":** Author interview with Bruce Sewell, March 2018.

181 **An NBC survey:** Devlin Barrett, "Americans Divided over Apple's Phone Privacy Fight, WSJ/NBC Poll Shows," *Wall Street Journal*, March 9, 2016, accessed September 14, 2018, www.wsj.com/articles/americans-divided-over-apples-phone-privacy-fight-wsj -nbc-poll-shows-1457499601.

181 **The United Nations voiced:** Buster Hein, "U.N. Backs Apple, Calls Encryption Fundamental to Freedom," Cult of Mac, March 3, 2016, accessed September 14, 2018, www.cultofmac.com/415765/u-n-backs-apple-calls-encryption-fundamental -to-the-exercise-of-freedom/.

181 **"there is no place":** Eugene Scott, "Comey: 'There Is No Such Thing as Absolute

Privacy in America,'" CNN, March 9, 2017, accessed September 14, 2018, www.cnn
.com/2017/03/08/politics/james-comey-privacy-cybersecurity/index.html.

182 **Lynch essentially accused:** SecureWorld News Team, "U.S. Attorney General Loretta Lynch Has Strong Words for Apple at RSA 2016," Cybersecurity Conferences
& News, March 9, 2016, accessed September 14, 2018, www.secureworldexpo.com
/industry-news/rsa-2016-us-attorney-general-loretta-e-lynch-has-strong-words
-apple-0.

182 **"Nothing could be further":** Author interview with Bruce Sewell, March 2018.

182 **Cook was preparing:** Author interview with Bruce Sewell, March 2018.

182 **The FBI said it had:** Matt Zapotosky, "FBI Has Accessed San Bernardino Shooter's
Phone Without Apple's Help," *Washington Post,* March 28, 2016, accessed September
14, 2018, www.washingtonpost.com/world/national-security/fbi-has-accessed-san
-bernardino-shooters-phone-without-apples-help/2016/03/28/e593a0e2-f52b-11e5
-9804-537defcc3cf6_story.html?utm_term=.7b37b9cb0d2e.

182 **It was later revealed:** Ellen Nakashima, "FBI Paid Professional Hackers One-Time
Fee to Crack San Bernardino iPhone," *Washington Post,* April 12, 2016, accessed
September 14, 2018, www.washingtonpost.com/world/national-security/fbi-paid
-professional-hackers-one-time-fee-to-crack-san-bernardino-iphone/2016/04/12
/5397814a-00de-11e6-9d36-33d198ea26c5_story.html?utm_term=.f8234cc590a4.

182 **At a Senate Judiciary hearing:** "Senator Reveals That the FBI Paid $900,000 to
Hack into San Bernardino Killer's iPhone," CNBC, May 8, 2017, accessed September
14, 2018, www.cnbc.com/2017/05/05/dianne-feinstein-reveals-fbi-paid-900000-to
-hack-into-killers-iphone.html.

182 **Officials had previously admitted:** Luke Dormehl, "FBI Found No New Information on San Bernardino Shooter's iPhone," Cult of Mac, April 20, 2016, accessed
September 14, 2018, www.cultofmac.com/424064/fbi-found-no-new-information
-on-san-bernardino-shooters-iphone/.

183 **"Fortunately, internet users":** "FBI Backs Down After Public Outcry, Opens San
Bernardino iPhone Without Apple's Help After Repeatedly Claiming That Was Impossible," Fight for the Future, March 28, 2016, accessed September 14, 2018, www
.fightforthefuture.org/news/2016-03-28-fbi-backs-down-after-public-outcry
-opens-san/.

183 **"Tim was a little disappointed":** Author interview with Bruce Sewell, March 2018.

183 **"You won't see this":** William Mansell, "Apple Adds New Animoji, Battery Improvements with Latest iOS 11 Update," *Newsweek,* March 30, 2018, accessed September
14, 2018, www.newsweek.com/ios-113-update-whats-new-animoji-battery-issues
-privacy-ios-11-features-867630.

184 **The release of iOS 11.3:** Scott Detrow, "What Did Cambridge Analytica Do During
the 2016 Election?," NPR, March 20, 2018, accessed September 14, 2018, www.npr
.org/2018/03/20/595338116/what-did-cambridge-analytica-do-during-the-2016
-election.

184 **Facebook confirmed that:** Anthony Cuthbertson, "How to Find Out If Your Facebook Data Has Been Compromised," *Independent,* April 9, 2018, accessed September 14, 2018, www.independent.co.uk/life-style/gadgets-and-tech/news/facebook-cambridge-analytica-users-personal-data-how-to-find-out-information-shared-a8295836.html.

184 **"I think that this":** "Apple's Tim Cook Calls for More Regulations on Data Privacy," Bloomberg, March 24, 2018, accessed September 14, 2018, www.bloomberg.com/news/articles/2018-03-24/apple-s-tim-cook-calls-for-more-regulations-on-data-privacy.

184 **"I wouldn't be in this situation":** Peter Kafka, "Tim Cook Says Facebook Should Have Regulated Itself, but It's Too Late for That Now," Recode, March 28, 2018, accessed September 14, 2018, www.recode.net/2018/3/28/17172212/apple-facebook-revolution-tim-cook-interview-privacy-data-mark-zuckerberg.

Chapter 10: Doubling Down on Diversity

185 **"Even among celebrities":** Author interview with Greg Joswiak, March 2018.

185 **On October 30:** "Tim Cook Speaks Up," Bloomberg, October 30, 2014, accessed September 13, 2018, www.bloomberg.com/news/articles/2014-10-30/tim-cook-speaks-up.

186 **"Kids were being bullied":** "Tim Cook on Speaking Up for Equality," YouTube, posted by *The Late Show with Stephen Colbert,* September 16, 2015, accessed September 14, 2018, www.youtube.com/watch?v=ZEq1qwos0w4.

187 **"From one son of the South":** Bill Clinton, Twitter post, October 30, 2014, 10:15 a.m., https://twitter.com/billclinton/status/527871526637699073.

187 **"Inspirational words":** Richard Branson, Twitter post, October 30, 2014, 7:35 a.m., https://twitter.com/richardbranson/status/527831152137355264.

187 **"He's chief executive":** James B. Stewart, "The Coming Out of Apple's Tim Cook: 'This Will Resonate,'" *New York Times,* October 31, 2014, accessed September 14, 2018, www.nytimes.com/2014/10/31/technology/apple-chief-tim-cooks-coming-out-this-will-resonate.html.

187 **"Tim sacrificed his own privacy":** "Tim Cook Receives the HRC Visibility Award," YouTube, posted by Human Rights Campaign, October 3, 2015, accessed September 14, 2018, https://youtu.be/iHguhlFE_ik.

187 **"just changed America":** Mark Gongloff, "Tim Cook Just Changed America in a Way Steve Jobs Never Could," *Huffington Post,* November 5, 2014, accessed September 14, 2018, www.huffingtonpost.com/2014/10/30/tim-cook-coming-out-water_n_6075388.html.

187 **"I felt like I was witnessing":** Author interview with Lisa Jackson, March 2018.

187 **"Would have been more interesting":** David Lazarus, Twitter post, October 30, 2014, 10:48 a.m., https://twitter.com/Davidlaz/status/527879616963170304.

188 **"Samsung just announced":** David [Wolf], Twitter post, October 30, 2014, 11:04 a.m., https://twitter.com/WolfSnap/status/527883649153527808.

188 **"To put it another way":** Seth Fiegerman, "Tim Cook Just Publicly Declared He's Gay and Wall Street Doesn't Care," Mashable, October 30, 2014, accessed September 14, 2018, https://mashable.com/2014/10/30/tim-cook-gay-wall-street-doesnt-care/#S9sz _Qn8psq6.

188 **"It may seem strange":** "Investors Don't Care That Tim Cook Is Gay," CNBC, www .cnbc.com/2014/10/31/investors-dont-care-that-tim-cook-is-gaycommentary.html.

188 **Cook was awarded a Visibility Award:** "Tim Cook at the Human Rights Campaign Annual Gala," C-SPAN, October 3, 2015, accessed September 14, 2018, www .c-span.org/video/?328534-2/tim-cook-human-rights-campaign-annual-gala.

191 **Greg Joswiak, Apple's vice president:** Author interview with Greg Joswiak, March 2018.

191 **In light of his and Apple's achievements:** Tim Bradshaw and Richard Waters, "Person of the Year: Tim Cook of Apple," *Financial Times,* December 11, 2014, accessed September 14, 2018, www.ft.com/content/4064a6fe-7fd7-11e4-adff-00144feabdc0.

192 **"I thought it would be impossible":** Bradshaw and Waters, "Person of the Year: Tim Cook of Apple."

192 **In November 2013:** Tim Cook, "Workplace Equality Is Good for Business," *Wall Street Journal,* November 3, 2013, accessed September 14, 2018, www.wsj.com/articles /workplace-equality-is-good-for-business-1383522254.

193 **"the future of our company":** Christina Warren, "Exclusive: Tim Cook Says Lack of Diversity in Tech Is 'Our Fault,'" Mashable, June 8, 2015, accessed September 14, 2018, https://mashable.com/2015/06/08/tim-cook-apple-diversity-women-future/.

194 **"The world is intertwined":** Corey Williams, "Tim Cook Discusses Diversity, Inclusion with Students," *Auburn Plainsman,* April 6, 2017, accessed September 14, 2018, www.theplainsman.com/article/2017/04/tim-cook-discusses-diversity-inclusion -with-students.

194 **"diversity with a capital D":** Sam Colt, "Tim Cook Gave His Most In-Depth Interview to Date—Here's What He Said," Business Insider, September 20, 2014, accessed September 13, 2018, www.businessinsider.com/tim-cook-full-interview-with-charlie -rose-with-transcript-2014-9.

195 **"I'm convinced we're going to move":** Warren, "Exclusive: Tim Cook Says Lack of Diversity in Tech Is 'Our Fault.'"

195 **"I think the US will":** Williams, "Tim Cook Discusses Diversity, Inclusion with Students."

196 **"I think it's our fault":** Warren, "Exclusive: Tim Cook Says Lack of Diversity in Tech Is 'Our Fault.'"

196 **In September 2011:** Eric Slivka, "Apple's Eddy Cue Promoted to Senior Vice President for Internet Software and Services," MacRumors, September 1, 2011, accessed September 14, 2018, www.macrumors.com/2011/09/01/apples-eddy-cue-promoted -to-senior-vice-president-for-internet-software-and-services/.

196 **In 2014, he promoted:** Peter Burrows, "Apple Promotes Young Smith to Run Human Resources," Bloomberg, February 11, 2014, accessed September 14, 2018, www .bloomberg.com/news/articles/2014-02-11/apple-promotes-young-smith-to-run -human-resources.

196 **and in May 2017 promoted:** Buster Hein, "Apple Promotes Denise Young Smith to Lead Diversity Efforts," Cult of Mac, May 23, 2017, accessed September 14, 2018, www.cultofmac.com/482976/apple-promotes-denise-smith-lead-diversity -efforts/.

196 **That same year:** "Sue Wagner Joins Apple's Board of Directors," Apple, July 17, 2014, accessed September 14, 2018, www.apple.com/newsroom/2014/07/17Sue-Wagner -Joins-Apple-s-Board-of-Directors/.

196 **Cook also instituted an annual:** Megan Rose Dickey, "Apple Releases First Diversity Report Under New VP of Diversity and Inclusion," TechCrunch, November 9, 2017, accessed September 14, 2018, https://techcrunch.com/2017/11/09/apple-diversity -report-2017/.

197 **A snarky post on Quartz:** Dan Frommer, "All the Women on Stage at Apple Keynotes, Charted," Quartz, June 8, 2015, accessed September 14, 2018, https://qz.com /422340/all-the-women-on-stage-at-apple-keynotes-charted/.

197 **But after Cook said:** "Apple Brought More Women to the Front at WWDC This Year. But Are the Numbers Good Enough?," *Economic Times*, June 5, 2018, accessed September 14, 2018, https://economictimes.indiatimes.com/magazines/panache/apple -brings-more-women-to-the-front-at-wwdc-this-year-but-are-the-numbers-good -enough/articleshow/64461770.cms.

197 **In 2014, 70 percent of Apple's:** Josh Lowensohn, "Apple's First Diversity Report Shows Company to Be Mostly Male, White," The Verge, August 12, 2014, accessed September 14, 2018, www.theverge.com/2014/8/12/5949453/no-surprise-apple-is-very-white -very-male.

198 **"As CEO, I'm not satisfied":** Lowensohn, "Apple's First Diversity Report Shows Company to Be Mostly Male, White."

198 **In November 2017:** Juli Clover, "Apple Publishes New Diversity and Inclusion Report," MacRumors, November 9, 2017, accessed September 14, 2018, www.macrumors .com/2017/11/09/apple-diversity-inclusion-2017-report/.

198 **The proportion of underrepresented:** Caroline Cakebread, "Apple Reiterated Its Commitment to Diversity—but It Made Little Progress in the Last Year and Is Still Predominantly White and Male," Business Insider, November 9, 2017, accessed September 14, 2018, www.businessinsider.com/apple-releases-2017-diversity-report -showing-little-progress-2017-11.

198 **According to the report:** Tony Romm and Rani Molla, "Apple Is Hiring More Diverse Workers, but Its Total Shares of Women and Minorities Aren't Budging Much," Recode, November 9, 2017, accessed September 14, 2018, www.recode .net/2017/11/9/16628286/apple-2017-diversity-report-black-asian-white-latino -women-minority.

198 **Apple's leadership page:** "Apple Leadership," Apple, accessed September 14, 2018, www.apple.com/leadership/.

198 **But Apple said it was:** "Inclusion & Diversity," Apple, accessed September 14, 2018, www.apple.com/diversity/.

199 **Sadly, Apple actually beats:** Rani Molla, "Facebook's and Twitter's Executive Leadership Are Still the Most White Among Big Silicon Valley Companies," Recode, July 3, 2017, accessed September 14, 2018, www.recode.net/2017/7/3/15913360/diversity-tech-report-google-gender-race.

199 **After a long and successful career:** Chris Weller, "Apple's VP of Diversity Says '12 White, Blue-Eyed, Blonde Men in a Room' Can Be a Diverse Group | Markets Insider," Business Insider, October 11, 2017, accessed September 14, 2018, https://markets.businessinsider.com/news/stocks/apples-vp-diversity-12-white-men-can-be-diverse-group-2017-10-1003866971.

199 **The comment seemed to be:** Matthew Panzarino, "Apple Diversity Head Denise Young Smith Apologizes for Controversial Choice of Words at Summit," TechCrunch, October 13, 2017, accessed September 14, 2018, https://techcrunch.com/2017/10/13/apple-diversity-head-denise-young-smith-apologizes-for-controversial-choice-of-words-at-summit/.

200 **"The nominating committee":** Adam Satariano, "Apple Facing Criticism About Diversity Changes Bylaws," Bloomberg.com, January 6, 2014, accessed September 14, 2018, www.bloomberg.com/news/articles/2014-01-06/apple-facing-criticism-about-diversity-changes-bylaws.

200 **Twice at shareholder meetings:** Anders Keitz, "Apple Investors Reject Diversity Proposal," TheStreet, February 28, 2017, accessed September 14, 2018, www.thestreet.com/story/14019740/1/apple-investors-reject-diversity-proposal.html.

200 **"They were put on":** Jacob Kastrenakes, "Apple Shareholders Are Demanding More Diversity, but the Company Is Fighting Back," The Verge, February 15, 2017, accessed September 14, 2018, www.theverge.com/2017/2/15/14614740/apple-shareholder-diversity-proposal-opposition.

200 **The board said an accelerated:** Megan Rose Dickey, "Apple Shareholders Make Another Push to Increase Diversity at the Senior and Board Levels," TechCrunch, February 2, 2017, accessed September 14, 2018, https://techcrunch.com/2017/02/02/apple-shareholders-make-another-push-to-increase-diversity-at-the-senior-and-board-levels/.

201 **"Meaningful change takes time":** Buster Hein, "Apple Diversity Report Shows the Company Is Still White and Male," Cult of Mac, November 9, 2017, accessed September 14, 2018, www.cultofmac.com/513275/apple-diversity-report-shows-company-still-white-male/.

201 **Most of the future job growth:** "Statistics," National Girls Collaborative Project, accessed September 14, 2018, https://ngcproject.org/statistics.

201 **"The reality is":** Corey Williams, "Special to The Plainsman: Tim Cook on Diversity

at Auburn," *Auburn Plainsman*, April 6, 2017, accessed September 14, 2018, www
.theplainsman.com/article/2017/04/special-to-the-plainsman-tim-cook-on
-diversity-at-auburn.

201 **Apple launched the Product Integrity Inclusion:** Aldrin Calimlim, "Apple
Launches New $10,000 Scholarship Program as Part of Its Push for Diversity,"
AppAdvice, September 3, 2014, accessed September 14, 2018, https://appadvice.com
/appnn/2014/09/apple-launches-new-10000-scholarship-program-as-part-of-its
-push-for-diversity.

202 **In 2015, Apple donated:** "Thurgood Marshall College Fund Head Discusses $40
Million Apple 'Investment' in HBCUs," Thurgood Marshall College Fund, May 7, 2015,
accessed September 14, 2018, www.tmcf.org/tmcf-in-the-news/thurgood-marshall
-college-fund-head-discusses-40-million-apple-investment-in-hbcus/4168.

202 **"What differentiates this partnership":** Michael Lev-Ram, "Apple Commits More
Than $50 Million to Diversity Efforts," *Fortune*, March 10, 2015, accessed September
14, 2018, http://fortune.com/2015/03/10/apple-50-million-diversity/.

202 **Apple donated $10 million:** Micah Singleton, "Apple Donates over $50 Million to
Improve Diversity in Tech," The Verge, March 10, 2015, accessed September 14, 2018,
www.theverge.com/2015/3/10/8184241/apple-donates-50-million-diversity-in-tech.

202 **NCWIT is also funded:** Lev-Ram, "Apple Commits More than $50 Million to Di-
versity Efforts."

202 **Both Facebook and Google:** KPMG U.S. Facebook post, July 28, 2018, www.face
book.com/KPMGUS/photos/weve-teamed-up-with-girls-who-code-gwc-a-national
-non-profit-organization-workin/979141295592861/.

202 **teaches 3-D printing and fashion:** Made with Code | Google, accessed September
14, 2018, www.madewithcode.com/.

203 **"We wanted to create opportunities":** Singleton, "Apple Donates over $50 Million
to Improve Diversity in Tech."

203 **"We're trying to encourage":** Author interview with Deirdre O'Brien, March 2018.

203 **"I would not be":** Aaron Smith, "Tim Cook Says Diversity Is Key to Great Companies," CNNMoney, August 24, 2015, accessed September 14, 2018, http://money.cnn
.com/2015/08/24/technology/tim-cook-apple-diversity/index.html.

203 **Apple's commitment to education:** "ConnectED," National Archives and Records
Administration, accessed September 14, 2018, https://obamawhitehouse.archives
.gov/issues/education/k-12/connected.

203 **Launched under President Obama:** "FACT SHEET: Opportunity for All—Answering
the President's Call to Enrich American Education Through ConnectED," National Archives and Records Administration, February 4, 2014, accessed September 14, 2018,
https://obamawhitehouse.archives.gov/the-press-office/2014/02/04/fact-sheet
-opportunity-all-answering-president-s-call-enrich-american-ed.

203 **This is a huge achievement:** Education Superhighway, "2017 State of the States:
Fulfilling Our Promise to America's Students," September 2017, https://s3-us-west-1.

amazonaws.com/esh-sots-pdfs/educationsuperhighway_2017_state_of_the_states.pdf.

204 **"These kids are born"**: "Exclusive: Apple CEO Tim Cook Talks Classroom Tech Initiative," ABC News, September 14, 2016, accessed September 14, 2018, https://abcnews.go.com/GMA/video/exclusive-apple-ceo-tim-cook-talks-classroom-tech-42072293.

204 **Apple under Cook has launched**: "Everyone Can Code," Apple, accessed September 14, 2018, www.apple.com/everyone-can-code/.

204 **"If I were a French student"**: *Konbini*, Facebook post, www.facebook.com/konbinifr/videos/10155995633024276/.

204 **"I want America to be"**: Shirin Ghaffary, "Full Audio: Our Extended, Uncut Interview with Apple CEO Tim Cook," Recode, April 7, 2018, accessed September 14, 2018, www.recode.net/2018/4/7/17210064/kara-swisher-tim-cook-chris-hayes-full-extended-uncut-interview-audio-podcast-download.

205 **In addition to the Everyone Can Code**: "Apple Celebrates Hour of Code at All Apple Stores," Apple, November 28, 2017, accessed September 14, 2018, www.apple.com/newsroom/2017/11/apple-celebrates-hour-of-code-at-all-apple-stores/.

205 **For those who can't make it**: www.apple.com/swift/playgrounds/.

205 **"At Apple we care deeply"**: Stefano Esposito and Mitch Dudek, "Apple Unveils New iPad in Chicago," *Chicago Sun-Times*, March 27, 2018, accessed September 14, 2018, https://chicago.suntimes.com/business/apple-unveils-new-ipad-in-chicago/.

206 **"We've fundamentally concluded"**: Aamer Madhani, "Apple CEO Tim Cook Wants to Teach Every Chicago Public School Student to Code," *USA Today*, December 12, 2017, accessed September 14, 2018, www.usatoday.com/story/news/2017/12/12/apple-teach-every-chicago-public-school-student-code/942609001/.

206 **"I suspect that tech"**: Madhani, "Apple CEO Tim Cook Wants to Teach Every Chicago Public School Student to Code."

206 **As new, younger people**: "Inclusion & Diversity," Apple, accessed September 14, 2018, www.apple.com/diversity/.

206 **"We want everyone"**: "Accessibility," Apple, accessed September 14, 2018, www.apple.com/accessibility/.

207 **"People with disabilities often"**: "Tim Cook Receiving the IQLA Lifetime Achievement Award," YouTube, posted by Auburn University, December 14, 2013, accessed October 4, 2018, youtube.com/watch?v=dNEafGCf-kw.

207 **"We design our products"**: "Tim Cook Receiving the IQLA Lifetime Achievement Award."

208 **Apple has promoted Global**: "Apple Brings Everyone Can Code to Schools Serving Blind and Deaf Students Nationwide," Apple, May 17, 2018, www.apple.com/newsroom/2018/05/apple-brings-everyone-can-code-to-schools-serving-blind-and-deaf-students/.

208 **"We're thrilled to kick off"**: "Apple Brings Everyone Can Code to Schools Serving Blind and Deaf Students Nationwide."

209 **"Apple's products are intuitive"**: Bill Holton, "Apple Receives AFB's Prestigious Helen Keller Achievement Award," *AccessWorld Magazine*, June 2015, www.afb .org/afbpress/pubnew.asp?DocID=aw160602.

209 **"We see accessibility"**: Steven Aquino, "When It Comes to Accessibility, Apple Continues to Lead in Awareness and Innovation," TechCrunch, May 19, 2016, https: //techcrunch.com/2016/05/19/when-it-comes-to-accessibility-apple-continues-to -lead-in-awareness-and-innovation/.

209 **"Consider someone with"**: Aquino, "When It Comes to Accessibility, Apple Continues to Lead in Awareness and Innovation."

209 **One of the individuals featured**: "Apple—Accessibility—Sady," YouTube, posted by Apple, October 27, 2016, accessed October 4, 2018, www.youtube.com/watch?v =XB4cjbYywqg.

209 **"That experience was the best"**: Sady Paulson, "Late Post (Apple)," *Sady Paulson* (blog), June 22, 2017, https://sadypaulson.com/2017/06/22/late-post-apple/.

210 **"Apple is founded"**: "Coffee with Tim Cook CEO of Apple," YouTube, posted by Accessible Hollywood, May 17, 2017, accessed October 4, 2018, www.youtube.com /watch?v=58ZZFUDIM0g&feature=youtu.be.

Chapter 11: Robot Cars and the Future of Apple

211 **On August 2, 2018:** Sara Salinas, "Apple Hangs Onto Its Historic $1 Trillion Market Cap," CNBC, August 2, 2018, www.cnbc.com/2018/08/02/apple-hits-1-trillion-in -market-value.html.

211 **Some experts attribute:** Tarun Pathak, "iPhone X Drove Apple's 'Revenue Super Cycle,'" Counterpoint, September 10, 2018, www.counterpointresearch.com/iphone -x-drove-apples-revenue-super-cycle/.

211 **"it's not the most important measure":** "Here's the Memo Apple CEO Tim Cook Sent to Employees After Hitting $1 Trillion," CNBC, August 3, 2018, www.cnbc.com /2018/08/03/apple-ceo-calls-1-trillion-value-a-milestone-but-not-a-focus.html.

213 **Logistics and transportation:** Zoë Henry, "Top 5 Industries (by Revenue) on the 2017 Inc. 5000," *Inc.*, August 22, 2017, accessed September 14, 2018, www.inc.com /zoe-henry/inc5000-2017-5-biggest-industries-revenue.html.

213 **"Apple is currently developing":** Matt Egan, "Apple Accused of Stealing Employees from Battery Maker," CNNMoney, February 19, 2015, accessed September 14, 2018, http://money.cnn.com/2015/02/19/technology/apple-stealing-employees-lawsuit/.

213 **"We believe the rumors":** Carl C. Icahn, "Carl Icahn Issues Open Letter to Tim Cook," Carlicahn.com, May 18, 2015, accessed September 14, 2018, http://carlicahn .com/carl-icahn-issues-open-letter-to-tim-cook/.

214 **Tony Fadell, former head:** Doug Bolton, "Steve Jobs Wanted to Make an Apple Car in 2008, Former Colleague Says," *Independent*, November 5, 2015, accessed September 14, 2018, www.independent.co.uk/life-style/gadgets-and-tech/news/steve-jobs -apple-car-2008-tony-fadell-a6722581.html.

214 **"A car has batteries":** Dawn Chmielewski, "Steve Jobs Tinkered with the Idea of an Apple Car the Year After the iPhone Premiered," Recode, November 4, 2015, accessed September 14, 2018, www.recode.net/2015/11/4/11620350/steve-jobs-tinkered-with-the-idea-of-an-apple-car-the-year-after-the.

215 **"They have hired people":** "All Charged Up in Berlin," *Handelsblatt Global*, November 30, 2015, accessed September 14, 2018, https://global.handelsblatt.com/companies/all-charged-up-in-berlin-316503.

215 **Zadesky's plan was:** Daisuke Wakabayashi, "Apple Scales Back Its Ambitions for a Self-Driving Car," *New York Times*, August 22, 2017, accessed September 14, 2018, www.nytimes.com/2017/08/22/technology/apple-self-driving-car.html.

215 **More than one hundred:** Mark Gurman and Alex Webb, "How Apple Scaled Back Its Titanic Plan to Take on Detroit," Bloomberg, October 17, 2016, accessed September 14, 2018, www.bloomberg.com/news/articles/2016-10-17/how-apple-scaled-back-its-titanic-plan-to-take-on-detroit.

216 **Cook spoke about Project Titan:** Alex Webb and Emily Chang, "Tim Cook Says Apple Focused on Autonomous Systems in Cars Push," Bloomberg, June 13, 2017, accessed September 14, 2018, www.bloomberg.com/news/articles/2017-06-13/cook-says-apple-is-focusing-on-making-an-autonomous-car-system.

216 **Employees familiar with the project:** Wakabayashi, "Apple Scales Back Its Ambitions for a Self-Driving Car."

217 **"the greatest failure":** Author interview with Horace Dediu, March 2018.

217 **Dediu told the story:** Author interview with Horace Dediu, March 2018.

218 **It was "a logistical nightmare":** Author interview with Apple employee, April 2018.

218 **In April 2006, Jobs announced:** "Steve Jobs' City Council Visit in 2006," YouTube, posted by City of Cupertino, April 18, 2016, accessed September 14, 2018, www.youtube.com/watch?v=XH7HcWQKxns.

218 **In June 2011:** "Steve Jobs Presents to the Cupertino City Council (6/7/11)," YouTube, posted by City of Cupertino, June 7, 2011, accessed September 14, 2018, www.youtube.com/watch?v=gtuz5OmOh_M.

218 **Today, that address is:** Michael Steeber, "Apple Marks Completion of New Campus with First Corporate Address Change Since 1993," 9to5Mac, February 17, 2018, accessed September 14, 2018, https://9to5mac.com/2018/02/16/apple-new-campus-corporate-address-one-apple-park-way/.

219 **"It's a little like a spaceship":** Alexia Tsotsis, "Jobs to Cupertino: We Want a Spaceship-Shaped, 12K Capacity Building as Our New Apple Campus," TechCrunch, June 7, 2011, accessed September 14, 2018, https://techcrunch.com/2011/06/07/steve-jobs-cupertino/.

219 **"from another planet":** Walter Isaacson, *Steve Jobs* (New York: Simon & Schuster, 2011), 570.

219 **"There's not a single straight piece":** Tsotsis, "Jobs to Cupertino: We Want a Spaceship-Shaped, 12K Capacity Building as Our New Apple Campus."

219 **"We do have a shot":** Tsotsis, "Jobs to Cupertino: We Want a Spaceship-Shaped, 12K Capacity Building as Our New Apple Campus."

219 **The company's plans were grandiose:** Steven Levy, "Apple's New Campus: An Exclusive Look Inside the Mothership," *Wired*, September 6, 2018, accessed September 14, 2018, www.wired.com/2017/05/apple-park-new-silicon-valley-campus/#slide-x.

220 **"He knew exactly what timber":** Levy, "Apple's New Campus: An Exclusive Look Inside the Mothership."

220 **"Steve means so much":** Irina Ivanova, "Apple iPhone 8, iPhone X, Watch Unveiled: As It Happened," CBS News, September 12, 2017, accessed September 14, 2018, www.cbsnews.com/news/iphone-8-release-apple-event-as-it-happened.

220 **"Maybe it should be called":** Stephen Fry, "When Stephen Fry Met Jony Ive: The Self-Confessed Tech Geek Talks to Apple's Newly Promoted Chief Design Officer," *Telegraph*, May 26, 2015, accessed September 14, 2018, www.telegraph.co.uk/tech nology/apple/11628710/When-Stephen-Fry-met-Jony-Ive-the-self-confessed-fanboi -meets-Apples-newly-promoted-chief-design-officer.html.

221 **Within weeks, there were reports:** Max A. Cherney, "People Are Walking into Glass at the New Apple Headquarters," MarketWatch, February 18, 2018, accessed September 14, 2018, www.marketwatch.com/story/people-are-walking-into-glass-at -the-new-apple-headquarters-2018-02-15.

221 **"The best, smartest designers":** Levy, "Apple's New Campus: An Exclusive Look Inside the Mothership."

221 **"Is Apple going to make":** Alissa Walker, "Steve Goes to the Mayor (Again)," *A Walker in LA* (blog), June 8, 2011, accessed September 14, 2018, www.awalkerinla .com/2011/06/08/steve-goes-to-the-mayor-again/.

222 **"Building campuses on":** Allison Arieff, "One Thing Silicon Valley Can't Seem to Fix," *New York Times*, July 8, 2017, accessed September 14, 2018, www.nytimes.com /2017/07/08/opinion/sunday/silicon-valley-architecture-campus.html.

222 **"You're never going":** Levy, "Apple's New Campus: An Exclusive Look Inside the Mothership."

223 **Mashable editor Lance Ulanoff:** Lance Ulanoff, "Inside Apple's Perfectionism Machine," Mashable, October 28, 2015, accessed September 14, 2018, https://mashable .com/2015/10/28/apple-phil-schiller-mac/#lxgN3Cweqkqr.

223 **In his book *Creativity Inc.*:** Philip Elmer-Dewitt, "What Architects Don't Get About Steve Jobs' Spaceship," *Fortune*, August 5, 2014, accessed September 14, 2018, http:// fortune.com/2014/08/05/what-architects-dont-get-about-steve-jobs-spaceship/.

224 **"Sometimes, we're having":** Author interview with Greg Joswiak, March 2018.

225 **"The first iPhone":** Seth Fiegerman, "iPhone X Features: 10 Things You Need to Know," CNNMoney, September 20, 2017, accessed September 14, 2018, https://money .cnn.com/2017/09/12/technology/gadgets/iphone-x-features/index.html.

225 **"the future of the smartphone":** "The Future Is Here: iPhone X," Apple, September 12, 2017, accessed September 14, 2018, www.apple.com/newsroom/2017/09/the -future-is-here-iphone-x/.

225 **It was the first iPhone:** "CORRECTED-UPDATE 5-Apple Unveils iPhone X in Major Product Launch," Reuters, September 12, 2017, accessed September 14, 2018, www .reuters.com/article/apple-iphone/corrected-update-5-apple-unveils-iphone-x-in -major-product-launch-idUSL2N1LT1BA.

226 **Even those who welcomed:** Macworld Staff, "iPhone X: Everything You Need to Know About Apple's Top-of-the-Line Smartphone," *Macworld,* December 1, 2017, accessed September 14, 2018, www.macworld.com/article/3222743/apple-phone /iphone-x-specs-features-release-date.html.

226 **"In terms of the way we price":** Catherine Clifford, "Apple CEO Tim Cook on $999 New iPhone X: 'We're Not Trying to Charge the Highest Price We Could Get,'" CNBC, November 3, 2017, accessed September 14, 2018, www.cnbc.com/2017/11/03/tim -cook-buying-a-999-iphone-x-is-like-buying-high-quality-coffee.html.

226 **Orders for the iPhone X:** Neil Hughes, "Notes of Interest from Apple's Q4 2017 Conference Call," AppleInsider, November 2, 2017, accessed September 14, 2018, https://appleinsider.com/articles/17/11/02/notes-of-interest-from-apples-q4-2017 -conference-call.

227 **"It's one of those things":** Yoni Heisler, "iPhone X Is a Flop? Actually, It's Apple's Best-Selling iPhone Model," BGR, May 2, 2018, accessed September 14, 2018, https:// bgr.com/2018/05/02/apple-iphone-x-sales-q2-2018-earnings/.

Chapter 12: Apple's Best CEO?

229 **"a product guy":** Author interview with Horace Dediu, March 2018.

230 **"We still think":** Author interview with Greg Joswiak, March 2018.

231 **"There's no chance":** Joel Hruska, "Ballmer: iPhone Has 'No Chance' of Gaining Significant Market Share," Ars Technica, April 30, 2007, accessed September 14, 2018, https://arstechnica.com/information-technology/2007/04/ballmer-says-iphone -has-no-chance-to-gain-significant-market-share/.

231 **"People forget that":** Author interview with Greg Joswiak, March 2018.

232 **Apple is estimated:** Tyler Lee, "Analyst Estimates 46 Million Apple Watch Units Sold to Date," Ubergizmo, May 4, 2018, accessed September 14, 2018, www.uber gizmo.com/2018/05/46-million-apple-watch-units-sold/.

233 **"Have you innovated today?":** Author interview with Greg Joswiak, March 2018.

234 **"Slowly, brands are waking":** Patrick Quinlan, "The Next Big Corporate Trend? Actually Having Ethics," Recode, July 20, 2017, accessed October 1, 2018, www.recode .net/2017/7/20/15987194/corporate-ethics-values-proactive-transformation -compliance-megatrend.

235 **If Lisa Jackson's estimates:** Author interview with Lisa Jackson, March 2018.

236 **more than 11.7 million workers:** Apple Inc., *Supplier Responsibility 2017 Progress Report,* www.apple.com/supplier-responsibility/pdf/Apple_SR_2017_Progress_Report.pdf.

236 **In March 2018:** Fred Imbert and Gina Francolla, "Facebook's $100 Billion-Plus Rout Is the Biggest Loss in Stock Market History," CNBC, July 27, 2018, accessed Septem-

ber 14, 2018, www.cnbc.com/2018/07/26/facebook-on-pace-for-biggest-one-day-loss-in-value-for-any-company-sin.html.

237 **Progress is slow:** "Inclusion & Diversity," Apple, accessed September 14, 2018, www.apple.com/diversity.

237 **"I don't think business":** Karen Gilchrist, "Apple's Tim Cook Shares a Rule That Leaders Should Live By," CNBC, June 26, 2018, accessed September 14, 2018, www.cnbc.com/2018/06/26/apple-ceo-tim-cook-advice-for-leaders-on-speaking-out.html.

Index

PENGUIN PARTNERSHIPS

Penguin Partnerships is the Creative Sales and Promotions team at Penguin Random House. We have a long history of working with clients on a wide variety of briefs, specializing in brand promotions, bespoke publishing and retail exclusives, plus corporate, entertainment and media partnerships.

We can respond quickly to briefs and specialize in repurposing books and content for sales promotions, for use as incentives and retail exclusives as well as creating content for new books in collaboration with our partners as part of branded book relationships.

Equally if you'd simply like to buy a bulk quantity of one of our existing books at a special discount, we can help with that too. Our books can make excellent corporate or employee gifts.

Special editions, including personalized covers, excerpts of existing books or books with corporate logos can be created in large quantities for special needs.

We can work within your budget to deliver whatever you want, however you want it.

**For more information, please contact
salesenquiries@penguinrandomhouse.co.uk**